An Effective Rigid Body Math Model

A Synopsis for the Practitioner

Dave Knopp

First Edition - Spring 2022

Stellacore Corporation

Made possible by the work and legacies of
Euclid, Euler, Grassmann, Clifford, Hestenes.

An Effective Rigid Body Math Model

First edition, May 2022.
Document ID: t_v1r1p0-1-g5792367

ISBN 978-1-387-90794-6

Attonook
Publishing

Produced with the L$_Y$X document processor and L^{A}T$_E$X typesetting.

Abstract

This technical brief summarizes an elegant algebraic description of rigid body position and attitude along with analytical derivatives associated with body motions. The equations presented here offer a fully general representation that is entirely singularity-free. The expressions capture all six degrees of freedom associated to a physical body within a math model containing exactly six free parameters. A particularly pragmatic benefit is that the math model and *all* of its parameters have direct associations with physical characteristics that can be observed and/or measured *directly* by physical instrumentation and sensor systems. The expressions describing body position and attitude are necessarily non-linear, but have relatively simple forms on which simple numerical solution methods are exceptionally effective.

Contents

Contents

Nomenclature

B Biector 3D angle representing the attitude of the body. Specifically, it expresses the angle through which the reference frame coordinate system must be turned in order to arrive at alignment with the body frame. All bivector components are expressed with resepct to the reference coordinate frame.

H The bivector half-rotation angle defined to be half that of B (e.g. $H \equiv \frac{1}{2}B$)

t Vector representing translation of the body. Specifically, it expresses the position of the body frame origin as expressed in the reference coordinate frame.

x Vector expressed the reference frame coordinate system

y Vector expressed the body frame coordinate system

Acknowledgments

The author wishes to thank those contributing assistance and feedback for this production. A special thank you goes to Scott Graybill and Vinod Khare for preliminary discussions sparking the incentive to get started and for providing ongoing motivation to reach completion. Thanks to Joe Bima for reviewing draft material and offering encouragement.

Part I.

Welcome

1. Preface

1.1. Purpose

The creation of this technical note is an unabashed attempt to promote the use of geometric algebra (GA) for representing rigid body state and motion, and the use of bivectors for representing rotation in particular. Both are an effective methodology for representing and describing rigid body position and attitude for both static and kinematic situations.

For those already sold on the use of GA, this technical note is intended to offer a useful "quick reference" for the various rigid body state and motion relationships.

1.2. Intended Audience

The content and presentation are directed toward individuals who are developing mathematical models or software implementations that address the state and motion of structurally rigid bodies - involving bodies from smart phones, to aircraft, to orbital satellites, to anywhere in between and beyond.

Readers are expected to have general math skills that cover classic algebra, practical use of vectors, and basic calculus, as well as some past experience with representing rigid body motion.

1.3. Caveats

This first edition is extremely rough. It is created in the spirit of what might be called "eXtreme Authoring" - a term that paraphrases the ideas and concepts behind a successful software development methodology nicely organized and reported in a series of "eXtreme programming" books by Kent Beck[1].

A key concept in eXtreme programming, and here now in eXtreme Authoring, is that of "rapid iteration". E.g. from the web page "extremeprogramming.org":

> Extreme Programming is successful because it stresses customer satisfaction. Instead of delivering everything you could possibly want on some date far in the future this process delivers the software you need as you need it.

This first edition represents gathering key ideas in one place. Improvements and polishing are left to next iterations sometime in the future when, and only if, they may be needed.

2. Introduction

The idea of a rigid body is commonly used to conceptualize physical objects, both to describe the situation of their static position and alignment in space and as well as their kinematic and/or dynamic movement through space and over time.

2.1. Rigid Bodies as High-Tech "Rocks"

The use and importance of rigid bodies predates the extent of our knowledge. Certainly the roots of "rigid body significance" exist at least as far back as some evolutionary ancestor learning to hold a rock in the right position and attitude to become an effective tool.

In the modern world, it's the same thing, except that we use a different and more elaborate set of tools. While our tools could be anything from a supersonic aircraft to a remotely controlled robotic manipulator arm holding a rock, it is still fundamentally important to be able to know and express object positions and attitudes in space.

2.2. Motivation

The concept of a rigid body is nearly ubiquitous in technical disciplines from engineering, transportation and navigation to describing the configuration of planets in astronomy, to expressing the position and attitude of a smart phone when taking a photograph.

A modern trend is toward devices that are loaded with sensors and which interact with their environments - e.g. self-driving vehicles, augmented and virtual reality systems, simulation and operations modules for digital twins and so on.

Many of these products and assets include various structural members and components that are often instrumented with a cornucopia of sensors, from tiny MEMS motion-sensing devices and various digital camera imaging systems, to sophisticated vehicle navigation and robot command and control systems. The various sensors, as well as components of the bodies on which they are mounted, are often treated as rigid bodies.

These overall systems and products typically involve software that is required for them to operate correctly and appear to be "smart". For many of these systems, the software applications must model the relative spatial configuration of individual component structures and/or motion of the overall ensemble within its environment.

2.3. Background

A rigid body is a body for which deformation is negligible in the context of the application of interest. In mathematical terminology, the distance between any two arbitrary points within the body remains constant. Because of this stability, a coordinate system can be attached to any arbitrary point on the body and be sufficient to represent the position and attitude of the entire body.

In mathematical terms, there is no preferred point of origin or preference in direction on a rigid body, and a coordinate system can be attached to the body at any position and in any attitude. In practice, of course, there is often some physically meaningful and/or convenient configuration for the coordinate frame. The mathematics presented herein are entirely general and work with whatever choice the practitioner decides.

2.3.1. Various Approaches

There is a very large number of mathematical approaches to describe the rigid body position and attitude for static configurations as well as for kinematics and dynamics settings.

These range from methods and techniques established during the development of classical mechanics several hundred years ago, to more relatively modern treatments based on advanced mathematical theories (cf. [10]). The approach presented herein is a kind of modern rediscovery of older methods that combines the best of each.

Serious mathematical modeling of rigid bodies began to take shape several centuries ago with the advent of algebra and calculus. Various modern techniques have been added onto this work, many of which arose from early work in quantum mechanics, mathematical developments in group theory, and work in differential geometry. More recent work in computer vision has popularized a number of other mathematical techniques.

Overall, the most practical current approach for rigid body modeling seems to revolve around a rediscovery and modernization of an older methodology - specifically, an integration of the mathematical formulations of Grassmann and Clifford, applied to the theories of Euler, capturing the geometric insights of Euclid as these have been reworked by Hestenes [5] to provide an integrated, modern, cohesive, and powerful mathematical language for classical mechanics [4].

2.3.2. Body Freedoms in 3D Space

The position of a rigid body in 3D space is associated with a single translation, comprising 3 physical degrees of freedom (3 dof), and a single rotation action, comprising another 3 physical dof. This means it requires six independent numeric values to specify the general geometric state of the body. For a moving body, each one of these 6 dof may be associated to a function having a value that changes in time.

Since there are 6 physical dof, any math model used to express state of the body must have at least 6 free parameters (or 6 functions of time for kinematic considerations).

2.3.3. Mathematical Parameters

Many mathematical approaches to rigid body attitude and motion utilize "over-parameterized" formulations such as the use of a rotation matrix, and quanternions.

The over-parameterization approach is fraught with danger. It artificially introduces extra mathematical dimensions into the problem and generally creates considerable inconvenience, numeric tribulations, and if not implemented with great care, can produce wrong answers.

Although over parameterized models may *appear* to make a problem "more linear", in general such an approach introduces a cornucopia of problems, additional new requirements, and/or various subtle and generally undesirable side effects.

The classic and well known minimum-parameter approach involves the use of Euler Angles[8] and the modified permutation known as "Tait-Bryan" angles[7]. Although these are minimum parameter representations for rotation, they (unnecessarily) introduce mathematical singularities associated with the physical phenomenon commonly known as "gimbals lock"[12].

2.3.4. A New "Old" Approach

The mathematical formulation summarized herein provides an elegant approach that *has minimum degrees of freedom* (dof) while being *free from singularities*. It is a *highly practical* mathematical formulation for representing *completely general* rigid body position and attitude (and motion) with mathematical expressions that have *direct physical interpretations*, and are *amenable to analytical analysis* as well as efficient numeric solution. These features are worth summarizing explicitly.

- Minimum degrees of freedom

- Free of singularities

- Entirely general case (no restrictions on applicability)

- Direct physical associations (to measurements and sensors)

- Analytically approachable (mesh with calculus - e.g. dynamics and kinematics)

- Practical (get the job done and easy to code into software)

- Coordinate-free representation - equations hold in any coordinate system framework

This last item is particularly powerful. It means that expressions can be formed, manipulated and interpreted independently from any particular underlying coordinate system convention (e.g. formulations are valid in curvilinear systems as well as rectilinear ones). However, this level of generality is beyond present purposes which focus on explaining practical rigid body concepts in the context of familiar Cartesian coordinate system frameworks.

3. Preliminaries and Context

3.1. Conventions

It is useful to establish several conventions. It should be noted that conventions certainly are not set in stone and are a matter of choice and preference. The following are used herein.

- Precedence of Operations - here using "translate then rotate" (compared with e.g. "rotated then translate" or "hyper rotate"). Ref. Appendix C.1.

- Relationship Order - here using "Into with respect to From" convention. E.g. "BwA" means coordinate frame "B" expressed wrt coordinate frame "A").

- State (aka pose, orientation) - used here to mean full 6 dof configuration addressing *both* body position and body alignment.

- Chirality - assuming dextral vector spaces. E.g. basis vectors, e_1, e_2, e_3 form a right handed triad when considered in cyclic order - i.e. from e_i to e_j and then to e_k.

- Cyclic Order - three indices i, j, k, are assumed to have values that are in cyclic order - i.e. $\{i, j, k\}$, are in the set $\{\{1, 2, 3\}, \{2, 3, 1\}, \{3, 1, 2\}\}$.

These conventions have been chosen, based on experience, as a particular convention that tends to integrate well with general engineering work, classical mathematics and general physics.

19

There are various other conventions that work nicely for particular situations. However, in the author's experience, the ones listed here are "overall" the most widely useful across multiple disciplines and application domains, for general modeling of practical systems that involve actual physical hardware and software.

3.2. Notation

Notation includes

Position (noun) - (aka location, offset) refers to where the body frame "is at"

Attitude (noun) - (aka pose, orientation, alignment) refers to how the body frame "is pointing"

Translate (verb) - (aka displace) refers to change in position of the physical body

Rotate (verb) - (aka turn) refers to change in attitude of the physical body

State (noun) - (aka orientation, pose) refers to the combination of position and attitude together. A note on this terminology: For kinematic and dynamic situations, the term "state" is often considered to also include an expression for velocities. E.g. considering "state" to include all of: position, attitude, a translation velocity, and a turning velocity. (sometimes accelerations might be included as well). In this interpretation, static case scenarios can be associated with states in which the velocity and acceleration constituents are all identically zero.

3.3. Ideal Point of Interest

When working with rigid bodies, it is common to consider the idea of an abstract "point of interest". The point being considered might have some physically tangible meaning, it might be arbitrary, or it could represent a variable. In any case, the point location can be represented effectively with a vector.

3.4. Transformation Concept

The rigid body can be subjected to physical changes in the form of displacements and/or rotations. These physical changes can be represented by mathematical expressions describing how the point of interest vector transforms.

The main emphasis here is that the mathematical expressions do not attempt to describe the physical body. Instead, they describe a mathematical operation that transforms (aka converts) *expressions*. The transformations change how *a same entity* (e.g. a vector) *is expressed differently* in the context of different coordinate frames.

In particular, the mathematical expressions herein are associated with how a vector entity (to a point of interest) is expressed relative to the context of two specific coordinate systems, a "reference frame" and a "body frame".

Descriptions and an example of details are presented in Appendix C.1.

Part II.

Rigid Body Transformations: Static Scenarios

4. State

A rigid body can be positioned at an arbitrary location in space by translating it along some line. Once at this new position, it can be set to an arbitrary attitude by turning it through some amount of rotation while constraining the rotation so that a specific "equatorial plane" remains unchanged.

Performing an arbitrary pure translation operation, followed by performing a pure rotation operation is sufficient to situate an arbitrary body at any general position and at any general attitude in space.

4.1. Terminology

In practices, there is a confusion of terminology when referring to this final specific result. Often it is called either "orientation" or "pose", with the other term being used to refer to attitude only. To remove this ambiguity for present work, the term "state" is used to mean *both* the position and attitude of a body.

In mathematical terms the translation action can be associated with a vector, while the rotation process can be expressed in multiple forms.

A main point is that a body state can be thought of as the result of a process, and that this situating process can be decomposed into a translation type displacement (aka an "offset") and a turning type displacement (aka a "rotation").

I.e. "State = Position + Rotation".

4.2. Degrees of Freedom

In three dimensional (3D) space, the translation involves 3 degrees of freedom (dof) and the rotation also involves 3 degrees of freedom. I.e. in 3D space, both aspects of the state have the same number dof[1].

---

[1]This equal partition in dof is unique to 3D space and is not true in other dimensions. In general, within an N-D space, translation has N dof, while rotation has $\frac{1}{2}N(N-1)$ dof. For example in 2D space, translation has 2 dof while rotation has only 1 dof. In 4D space, translation has 4 dof while rotation has 6 dof.

5. Transformation Components

Let the symbol, x, represent the vector to the point of interest as it is expressed in some arbitrary reference frame "R". Let the symbol, y, represent *the same point* but with expression as it occurs in the body frame "B".

The direction of the transformation is denoted herein with a notation that might be described as "Into (with respect to) From" which is abbreviated as "IwF". The case of describing the body coordinate frame state relative to a reference frame is therefore denoted as "BwR".

E.g. letting $\mathcal{T}_{BwR}$ represent the composite displacement and rotation transformation, the rigid body position and attitude can be associated with a mathematical transformation that operates on a point of interest,

$$y = \mathcal{T}_{BwR}\left(x\right)$$

There are two particularly common procedures associated with rigid body work. These are:

- Specify an expression for, $\mathcal{T}_{BwR}$, given physical configuration information about a rigid body.

- Infer physical configuration information given the behavior of the transformation expression, $\mathcal{T}_{BwR}$.

Using the methodology below the first of these becomes a trivial operation. The second is conceptually simple, but, in practice, requires solving a non-linear multivariate equation. However, the equation is amenable to simple numeric methods.

5.1. Translation

The displacement operation, $\mathcal{D}_{BwR}$, may be described by a single vector-valued parameter, t. The displacement translation is then described by a displacement operation that corresponds with vector subtraction,

$$\mathcal{D}_{BwR}\left(x, t\right) = x - t$$

This may be interpreted as follows.

Whereas "x" is interpreted as the location of the point relative to the domain (reference) frame origin, "$x - t$" can be interpreted as the point of interest now "expressed wrt to location t"[1].

Note that the quantity, $(x - t)$, is still expressed in the transform domain!! (e.g. for a BwR transformation, the difference vector, $x - t$, is a vector that has interpretation relative to coordinate frame "R")

For a pure translation (no rotation),

$$y = x - t \tag{5.1.1}$$

5.2. Rotation

The rotation operator, $\mathcal{R}_{BwR}$, can be represented in many (many, many) forms. The structure outlined here associates the *rotation operation* (not the attitude parameters) with an entity known as a "spinor" (aka "rotor").

A spinor is algebraically represented as the exponentiation of a bivector (aka "3D angle", or "pure vector quaternion"). The

---

[1]I.e. the displacement operator converts interpretation from an "absolute sense" to a "relative sense".

spinor may also be interpreted as an "exponential map" and is associated with group theory SE(3), the special subgroup of SO(3). The practical interpretation and details of this are explained in following sections and in Appendices A.1.6 and A.4.

The various rotation formulations are intimately related to each other (some are isomorphic others have practical equivalents over some restricted domain). However, the constructs of "geometric algebra" (GA) provide a particularly concise and elegant formulation.

In particular, GA expressions offer a *powerful physical interpretation* that directly associates *physical properties* and with *instrumentation/sensor data.*

Forgoing additional soap box evangelism about the power available from GA interpretations, it's time to dive into actual expressions below. A touch of supporting practical background material for GA is provided in the Appendices.

5.2.1. Half-Rotation Parameters

As described in Appendix B.1, a rotation may be associated with a bivector, B, that represents a full "3D angle". This bivector can be expressed in terms of three independent scalar parameters (ref. Appendix A.1.3) which correspond directly to the three degrees of freedom associated with physical rotation in 3D space.

In terms of the bivector scalar magnitude, $|B|$, and (planar) direction, $\hat{B}$, a pure rotation (no translation) may be expressed by the rotation transformation,

$$y = e^{\frac{1}{2}|B|\hat{B}} \left(x\right) e^{-\frac{1}{2}|B|\hat{B}}$$

This form of expression emphasizes that the rotation operation, via the arguments of the exponential function, include both the 1 dof of magnitude of the rotation (via $|B|$), and also the 2 dof of the direction of rotation (via $\hat{B}$).

5. Transformation Components

For algebraic work, it's often convenient to suppress the factor of one half, and work directly with the concept of a *half-rotation* (ref. Appendix B.2).

For example, introduce bivector, H, to represent the half-angle, and express it in terms of scalar magnitude parameter, ϑ, and bivector direction parameter, Θ, as

$$H \equiv \frac{1}{2}B$$
$$\vartheta \equiv |H|$$
$$\Theta \equiv \hat{H}$$

Here, the bivector, H, is a complete expression of the half-rotation angle. The scalar, ϑ, expresses the magnitude, or "amount of turning" for the (half) rotation (in radians). The unitary (unit-magnitude) bivector, Θ, encodes the direction of the half-rotation (represents the direction of the equatorial plane within 3D space).

In terms of the half-rotation angle, H, a pure rotation (no translation) can be expressed by the transformation relationship,

$$y = e^H x e^{-H} \tag{5.2.1}$$

The expression, H, is often used for numerical work and the magnitude/direction decomposition is often useful for interpretation and analysis. The expression, B, is often used for physical interpretation (e.g. as "the" attitude of an object).

In terms of these parameters, the rotation spinor may be expressed equally as,

$$e^H = e^{\vartheta\Theta} = e^{\frac{1}{2}B}$$

The exponential function is defined algebraically as expected. I.e.

$$e^H = \sum_{n=0}^{\infty} \frac{1}{n!} H^n = 1 + H + \frac{1}{2!} H^2 + \frac{1}{3!} H^3 + \dots$$

28

5. Transformation Components

As with other expressions herein, the algebraic notation should be interpreted in the general context of geometric algebra. For example, here, the powers of the bivector, H, are computed using the geometric product (ref Appendix A.3).

Since bivectors square to non positive values the exponential series is an alternating series. This property is useful for knowing bounds on truncation error when evaluating the exponential function via series expansions. It also leads to a generalization of Euler's identity,

$$e^H = e^{\vartheta \Theta} = \cos \vartheta + \Theta \sin \vartheta$$

In Euler's identity, the first term is a scalar entity and the second is a bivector entity. I.e. e^H is a spinor (ref. Appendix A.1.6). Therefore the exponential function (which is being used here as the rotation operator) can be decomposed into the product of two bivectors as illustrated in Appendix A.3.

The exponential term can also be decomposed into a product of two unit vectors, as illustrated in Appendix A.2.

6. Full Rigid Body Transformation

Using the half-angle rotation operator from above, and reintro-
ducing the body translation, a *completely general, singularity
free*, and *readily differentiable*, rigid body transformation can be
expressed, as

$$y = e^{H} (x - t) e^{-H} \qquad (6.0.1)$$

In mathematical terms, this is the composition of the pure
translation transformation from expression 5.1.1 followed by the
pure rotation transformation of expression 5.2.1.

The full rigid body transformation expression 6.0.1, represents
the complete geometry of a rigid body. It includes the body displacement and rotation in a single concise mathematical transformation expression. It captures all 6 dof in 6 (and only 6) free
parameters (the three scalar components t, e.g. $\{\tau_1, \tau_2, \tau_3\}$, and
the three scalar components of H, e.g. $\{\eta_1, \eta_2, \eta_3\}$).

6.1. Translation Parameters

The components of t, have the expected vector interpretation.
E.g. the body is moved τ_k units along the e_k axis. The τ_k
components represent the projection of this displacement vector
onto each of the coordinate frame axes.

This interpretation of translation motion can be considered to
be either sequential or simultaneous. I.e. interpretation of the
body moving directly along a line (a geodesic) associated with

vector t (e.g. directly from "tail" to "tip"), or interpretation of moving first by τ_i units along the e_i axis, then τ_j units along the e_j axis ($j \neq i$) and then τ_k units along the third, e_k, axis ($k \neq \{i, j\}$).

For displacement, the simultaneous and sequential interpretations both produce the same resulting position. This is because the mathematical operations of vector addition (subtraction) are commutative (e.g. $\tau_1 e_i + \tau_j e_j = \tau_j e_j + \tau_i e_i \, \forall i, j$).

6.2. Rotation Parameters

The components of H, have the interpretation analogous to the components of the displacement vector. Each of the η_k represent the projection of the body attitude bivector (rotation 3D angle) onto each of the three coordinate frame basis planes. E.g. having relative magnitudes that match the projection of the rotation "equator" onto each of the three coordinate planes, and having overall magnitudes in proportion to *one-half* the overall rotation amount. The η_k rotation components must be interpreted as simultaneous rotation parameters.

Note that the components of H, can also be considered in any order because the bivector components of H are commutative under bivector addition!! E.g. $\eta_i E_{jk} + \eta_j E_{ki} = \eta_j E_{ji} + \eta_i E_{jk}$ (for all $\{i, j, k\}$ in cyclic order). This means that there is no precedence or order among the three individual components of H, and that each of the η_k values stands equally on its own!

6.2.1. Simultaneous Rotation

The attitude parameter values, e.g. $\{\eta_1, \eta_2, \eta_3\}$, must all be considered together to represent the attitude of the body. There is no ordering associated with the parameters. Contrast this with the very different "sequential" specification of attitude described in the next section.

Another way to phrase this, is that the three components of H provide an "instantaneous" and "intrinsic" expression for (one-half) the attitude of the body. This specification of the (half-) attitude is *completely independent* of any special interpretations or implicit assumptions about some fictitious procedure or path of "getting there".

The interpretation and benefits of this simultaneous interpretation may be better understood by contrast to the extrinsic and procedural attitude representations discussed in the next section.

Additional benefits become apparent in the context of kinematic considerations addressed in Sections 7.3.1 and 7.3.2.

As an aside, note that this simultaneous rotation interpretation is *exactly* consistent with a *"strap down"* attitude measurement system (e.g. inertial measurement units, or use of computer vision/photogrammetric or surveying systems that derive attitude from simultaneous sightings).

This simultaneous association is often very useful in practice exactly because most bodies actually do "turn simultaneously" in all three dimensions!

6.2.2. Sequence of Rotations

By contrast to the simultaneous interpretation of angle components just described, there is a radically different "procedural" technique for describing body attitude as a sequence of three individual, mutually independent, rotation operations (e.g. the various Euler angle[8] conventions and Tait-Bryan angle[7] conventions).

It is important to note that the sequential rotation interpretation is an entirely *different physical concept* than the sequential interpretation addressed above. This section exists primarily to emphasize this contrast (although the sequential representation

6. Full Rigid Body Transformation

is also useful in practice such as for bodies physically constrained to move about mechanical axes[1]).

The sequential methodologies are based on a procedural/operational and *extrinsic* definition! It involves a process of "getting to" some desired attitude by following a implicit procedural "recipe". By contrast, the simultaneous convention inherent in the bivector angle representation, is a fundamentally *intrinsic* way to parameterize the attitude.

For comparison and contrast, the various sequential representations are readily expressed by a simple extension of expression 6.0.1. To obtain a sequential rotation model, define three individual rotation concepts (rather arbitrarily, but being sure that the three together span all possible rotations), then simply apply each successive independent rotation operation to the result of the previous one.

For example, a general sequential rotation process can be expressed as

$$y = e^{H_3} e^{H_2} e^{H_1} (x - t) e^{-H_1} e^{-H_2} e^{-H_3}$$

where the H_k are three suitable bivector half-angles in space (typically orthogonal to each other, but need not be). Each of these is of the form, $H_k = |H_k| \hat{H}_k$, etc. The $|H_k|$ is the free parameter and the $\hat{H}_k$ are a set of preferred *constant* (bivector) directions.

For example a simple "x-y-z", aka "roll-pitch-yaw", type rotation sequence can be expressed as,

$$y = e^{\eta_z E_{12}} e^{\eta_y E_{31}} e^{\eta_x E_{23}} (x - t) e^{-\eta_x E_{23}} e^{-\eta_y E_{31}} e^{-\eta_z E_{12}}$$

As required for 3D space, this sequential rotation approach still has three free parameters for rotation. However, it obvi-

---

[1]It is, for example, exactly the right model for a multiple gimbals, gyroscopic system.

ously involves a considerably more complex non-linear relationship between the free parameters compared with the more direct, simultaneous relationship in expression 6.0.1.

On the positive side, these sequential representations also have useful interpretations when relatively small magnitude rotations are involved. For example the situation of a pilot controlling an aircraft, or a tele-operator controlling a robot manipulator, in which small perturbations and incremental changes to attitude are most relevant.

The extra non-linearity involved with the sequential rotation method often lead to various implementation problems (e.g. much more complexity and/or difficulty with numeric solutions) and/or undesirable behaviors (e.g. introduction of singularities associated with "gimbals lock").

Notwithstanding the above, this model for attitude is consistent with, and particularly suitable for, attitude determination systems that involves gimbals technology. The (ordered) sequence of individual rotation operations corresponds with the nesting order of the physical gimbals. The mathematics correctly model the physical device operations including when they "lock".

As a result of the non-linear combination of free parameters, the individual rotation operations are (highly) non-commutative. The amount of error involved diminishes as the size of the individual sequence angles gets small. In this case the sequence of angle operations is "almost" order independent for small attitude changes. This "almost effect" often leads to the introduction of problems, and hamstrings the ability to detect them. Therefore, great care is generally required to ensure that sequential rotation model formulations are implemented correctly.

Part III.

Rigid Body Transformations: Kinematics and Derivatives

7. Motion

Here the word "kinematic" is intended in two contexts. The first is the common case to mean "dealing with moving objects". The second context might be phrased as "dealing with moving (numeric) solutions". Both situations make extensive use of the derivatives of the rigid body transformation relationships.

For example in the "moving body" interpretation, successive differentiation of the relationship in expression 6.0.1 provides explicit relationships for rigid body velocity and rigid body accelerations. The later relate directly to various forces and torques associated with dynamical considerations as well as phenomena that can be measured by physical instrumentation and sensors.

On the other hand, when computing solutions with numerical methods, expressions for derivatives provide a means to accelerate solutions such as via various descent methods that utilize function gradients and Laplacian/Hessian relationship. The derivatives also make it simple to assess solution stability and analyze first order sensitivities of configurations and solutions.

In short, being able (or at least having the option) to compute analytical derivatives is often very useful. The following sections explore differentiation of the canonical rigid body transformation relationship from expression 6.0.1.

7.1. Paths

The position and attitude of a static rigid body is associated with six degrees of freedom (and the corresponding six scalar parameters in the full transformation model of expression 6.0.1. For a body in general motion, each of the 6 freedoms and parameters changes as a function of time. This can be thought of as a path described by parametric equations within a 6D state space. I.e. the body states is expressed as a 6 dof function of a scalar sequencing parameter.

7.1.1. Parameter Functions

For a rigid body that changes position and/or attitude over time, the parameters expressing its 6 dof state are replaced by functions of time. E.g.

$$t = t(\sigma)$$
$$H = H(\sigma)$$

where the scalar parameter, σ, is sequencing index. Often this is some form of a time index such as number of clock ticks or a counter value, or sometimes an image exposure sequence number, or data record number, etc.

For the kinematic system, $t(\sigma)$, is a vector-value function of the scalar sequence index, σ, and $H(\sigma)$, is a bivector valued function of this same scalar index.

Although the following assumes these functional relationships, the excess baggage of the function notation is not needed, as long as it is remembered that t and H should be interpreted:

- For static cases: as *parameter values for static situations*

- For kinematic cases: as *parametric functions* of some sequence scalar, σ

7.1.2. Curve in State Space

The kinematic history of a rigid body can be conceptualized as a curvilinear path through 6D space. The coordinates of a point on this path can be indexed by the 6 scalar parameters, $\{\tau_1, \tau_2, \tau_3, \eta_1, \eta_2, \eta_3\}$, with each parameter being a scalar function of some sequencing index, σ. Note that the 6 individual scalar parameters are mutually orthogonal and independent of each other!

7.2. Displacement Motion

The position path followed by the origin of the body coordinate frame is described directly and entirely by the 3D translation vector, t, or more precisely by the vector parametric function, $t(\sigma)$.

For dynamical systems it's often convenient to associate the rigid body coordinate frame origin with the body center of mass. For other situations and applications it is more convenient to associate t with a point of reference or measurement (e.g. the location of a particular sensor, or a particular point of some structure).

7.2.1. Interpolation and Extrapolation

The rigid body transformation is linear in the translation parameter, t, which has three independent scalar component parameters, τ_k. It is therefore simple and easy to interpolate (or extrapolate) rigid body (coordinate frame) positions given a collection of known or measured locations.

For example, a simple quadratic model is often useful, such as

$$t(\Delta\sigma) = t_0 + t_1(\Delta\sigma) + t_2(\Delta\sigma)^2$$

This is continuously differentiable. Also, the second order nature is useful in kinematics and dynamics work for expressing accelerations related to various force models.

When working with discrete data, various forms of spline models can be particularly useful in preserving continuity of position values, velocities, and accelerations, across individual data samples.

7.3. Turning Motion

The attitude path followed by the alignment of the body coordinate frame is described directly and entirely by the 3D half-angle bivector, H, or more precisely by the continuously differentiable bivector parametric function, $H(\sigma)$.

7.3.1. Simultaneous Components

As noted for the static case (ref Section 6.2.1), the component parameters of H, i.e. the η_k scalar values, represent three distinct components of *one single rotation operation*. There is *no* order or precedence associated with the components. They all contribute equally and simultaneously to specification of the complete half-angle bivector, H.

For illustration, consider a body that is rotating uniformly and has attitude, H_1, at some sequence index (e.g. time), σ_1, and then later has attitude described by, H_2, at index, σ_2. Since the body is in uniform motion, it "rotates directly into attitude H_2 from H_1" along a geodesic path between the two attitudes.

In this illustration, all three component parameters, the η_k, have values that vary in lock-step with each other. This makes it very simple to interpolate (or extrapolate) body attitude at times, σ. E.g. the intermediate (or extended) attitude may be computed as

$$H(\sigma) = \frac{\sigma - \sigma_1}{\sigma_2 - \sigma_1}(H_2 - H_1) + H_1 \qquad (7.3.1)$$

In addition to illustrating the simultaneous interpretation of parameters (e.g. as a "direct path" or "geodesic path" of rotation), this is actually a useful interpolation formula as addressed further in the next section.

7.3.2. Interpolation and Extrapolation

The linear interpolation function in expression 7.3.1, corresponds with the "SLERP" interpolation method commonly used for animating rotations of objects in gaming and computer graphics applications (cf. [13]).

The "S" in SLERP stands for "spherical". However, note that interpolation formula 7.3.1 is *not* spherical in nature but is instead, a simple *linear* interpolation expression. The associated "spherical" nature is actually associated with the spinor, e^H, and not with the angle, H. This has profoundly useful practical implications.

The simplicity of the example interpolation model in expression 7.3.1 rightfully suggests that there is no reason to limit interpolation to a linear model!

A variety of higher order angle interpolation models can be easily employed to represent more realistic interpolation of physical motions[1]. I.e. instead of being restricted to piece-wise linear interpolation of rotations, other more sophisticated interpolations can be utilized that provide higher order smoothness

In particular, it is easy to implement interpolation models that have differential smoothness of varying degrees. This can be exceptionally helpful in satisfying physical considerations in kinematic and dynamical models (such as motion being consistent with environmental torques). For example, a simple quadratic

---

[1]Such as when a body is subjected to torques imparted by interaction with its environment.

model along the lines of

$$H\left(\Delta\sigma\right) = H_0 + H_1\left(\Delta\sigma\right) + H_2\left(\Delta\sigma\right)^2$$

provides rotation interpolation that is continuous to second order.

As is the case with the vector translation, various spline models are often useful for interpolating turning motions across discretely sampled data values in a manner that provides an appropriate order of physical smoothness and continuity.

8. Transformation Derivatives

In the following, a differentiation operation, wrt to some arbitrary assumed scalar, is indicated with two notations, either with an overdot notation where the argument for the differential operator is obvious, or with an overline where it better clarifies the scope of the differentiation operation. E.g. for some arbitrary scalar, ζ, and arbitrary entity Q,

$$\dot{Q} \equiv \overline{Q} \implies \frac{\partial Q}{\partial \zeta}$$

8.1. Body Motions

Expressions for rigid body transformation relationships can be differentiated with respect to an implicit kinematic sequence parameter, σ. When σ is interpreted as a time parameter (e.g. clock tics, data record sequence index, etc), the derivatives provide various "velocity" relationships.

Expressing changes in the position of a body is relatively simple because the rigid body transformation expression 6.0.1 is linear in the translation parameter, t. However, expressing changes in body attitude is more complex and involves several subtleties.

Velocities associated with turning motion can be expressed in multiple ways, including

Angular Velocity

Expressions involving the rate of change of the attitude

angle, B, or of the half-angle, H. I.e. expressions involving $\dot{B}$, or $\dot{H}$. Noting that $\dot{B} \equiv 2\dot{H}$. The angular velocity is a change in the angle expressing body attitude.

Rotational Velocity

This is a more abstract, but also very convenient, concept for body turning motions. It is related to the derivative of the spinor exponential function, $\overline{e^H}$, (ref below for detail). The rotational velocity is therefore a function of angular velocity, but it has distinctly different algebraic properties and interpretation.

Both types of turning velocities are useful in practice. For example, angular velocities are useful when working with angular position sensors such as gyroscopes, while rotational velocities are useful for expressing mathematical operations and working with sensors such as Coriolis-effect turning rate sensors[1]

8.1.1. Body Angular Velocity

The body's angular velocity is associated with change in the attitude angle parameter, B. The angular velocity (not to be confused with "rotational velocity" discussed below), is the rate of change in the attitude angle as a function of some scalar sequence parameter, σ. It is expressed simply as,

$$\dot{B} = \frac{\partial}{\partial \sigma} B$$

In a Cartesian frame,

$$\dot{B} = \dot{\beta}_1 E_{23} + \dot{\beta}_2 E_{31} + \dot{\beta}_3 E_{12}$$

---

[1]Such as MEMS sensors that use the Coriolis effect to infer rotation rates. They are often called "gyros"[sic], but this is a misnomer in that true gyroscopes measure angular positions and rates, not rotational positions and rates!

Note that this is twice the half-angle velocity,

$$\dot{B} = 2\dot{H}$$

Over time, the attitude angle, B, traces a path in a 3D bivector parameter space. The angular velocity is the bivector representing the instantaneous osculating circle of this path. In physical terms, it may be interpreted as an instantaneous equator of rotation.

8.1.2. Body Rotational Velocity

The exponential of the half-angle is a useful entity. Define the entity, R, in terms of the exponential function of the half-angle (ref. Appendix A.4 for detail on the exponential function of a bivector),

$$R \equiv e^{H} \qquad (8.1.1)$$

Here, R is an entity often called a "spinor" (or "rotor"). It is the sum of two constituents, a scalar part, and a bivector part, and therefore comprises four individual scalar components. E.g. in a classic Cartesian coordinate frame,

$$R = \rho_0 + \rho_1 E_{23} + \rho_2 E_{31} + \rho_3 E_{12}$$

Spinor, R, is unitary (since the argument in the exponent is a pure bivector). This unitary condition can be expressed with the algebraic constraint relationship,

$$|R| = \sqrt{RR^{\dagger}} = \sqrt{\sum_{\mu=0}^{4} \rho_\mu} = 1$$

Using basic GA relationships, it is easy to show that the spinor derivative, $\dot{R}$, can be expressed as a function of a bivector, Ω, such that

$$\dot{R} = \frac{1}{2}\Omega R \qquad (8.1.2)$$

This expression describes the rotational velocity (rate of change of the spinor, R).

Note that this relationship effectively incorporates the unitary constraint, so that the rotational velocity, $\dot{R}$, is expressed in terms of a rotational velocity bivector angle, Ω, which has only 3 dof and can be described uniquely and entirely with only three free parameters!

After multiplication from the right[2] by $2R^\dagger$, this can be interpreted as the defining equations for rotational velocity, Ω,

$$\Omega \equiv 2\dot{R}R^\dagger \qquad (8.1.3)$$

The bivector, Ω, is known as the "rotational velocity" (distinct from the angular velocity $\dot{H}$).

It is important to distinguish interpretations between the "rotational" and "angular" turning velocities. The rotational velocity, Ω, is a somewhat abstract "[constrained]spinor velocity" (aka a "spherical velocity") and not directly associated with changes in angles.

Rotational velocity, Ω, is more like a change in "perceived (observed) attitude" of the body. Angular velocity, $\dot{H}$ (or $\dot{B}$), is more like a change in "specification of (parametric) attitude".

The bivector representation of rotational velocity, Ω, is instrumental in describing and interpreting the various "fictitious" forces associated with rotating coordinate systems!!

For a rotating body, the angular velocity, $\dot{H}$ (or $\dot{B}$), is particularly relevant for expressing motion w.r.t. the reference frame, while the rotational velocity, Ω, tends to have significance when considering effects in the body frame. In one sense, angular velocities are often particularly useful for describing inertial motions, while rotational velocities are often particularly useful for

---

[2]Noting that, because R is unitary, the reverse and inverse are equal, i.e. $R^\dagger = R^{-1}$.

describing motions that are observed from "inside" turning reference frames.

The rotational velocity, Ω, may be expressed in terms of a function of the half-angle, H, and its angular velocity derivative, $\dot{H}$. This can be accomplished by substitution of defining relationship 8.1.1 into defining relationship 8.1.3, to produce an equivalent definition for rotational velocity, Ω, in terms of exponential functions of the half-angle, H, as

$$\Omega = 2\overline{e^{H}}e^{-H} \qquad (8.1.4)$$

The derivative of the exponential function can be expressed in terms of angular velocities, $\dot{H}$, and angular position, H, as described in Appendix D. E.g. Ω is a function of H and $\dot{H}$,

$$\Omega = \Omega\left(H, \dot{H}\right)$$

In principal (and in practice) this functional relationship can be inverted to express the angular velocity, $\dot{H}$, as a function of the rotational velocity, Ω, to obtain,

$$\dot{H} = \dot{H}\left(\Omega, H\right)$$

This result is described by Bortz with convention vector math in [2], while Boyle summarizes the relationships in terms of spinor and bivector quantities in [3].

8.1.3. Body Rotational Acceleration

The rotational acceleration is associated with the derivative of the rotational velocity.

Differentiation of the rotational velocity relationship equation 8.1.4 can be expressed as

$$\dot{\Omega} = \overline{2\dot{e^{H}}e^{-H}}$$

$$= \overline{2\dot{e^{H}}\left(e^{H}\right)^{\dagger}}$$

Applying the differentiation product rule, this may be expanded as

$$\dot\Omega = 2\overline{\overline{e^H}}\left(e^H\right)^\dagger + 2\overline{e^H}\overline{\left(e^H\right)^\dagger}$$
$$= 2\overline{\overline{e^H}}\left(e^H\right)^\dagger + 2\overline{e^H}\left(\overline{e^H}\right)^\dagger$$

which leads to the expression,

$$\dot\Omega = 2\overline{\overline{e^H}}\left(e^H\right)^\dagger + 2\left|\overline{e^H}\right|^2 \qquad (8.1.5)$$

The second term on the last line is simply the square of the spinor derivative that can be computed directly from equation 10.1.1 yet to come. However, the first term involves second order differentiation of the exponential function which is more involved (ref. Section 10.1.4).

8.2. Point Motions

8.2.1. Point Velocity in the Body Frame

Differentiation of the rigid body transformation may be addressed by applying the differentiation product rule to the canonical rigid body transformation expression 6.0.1, with the result expressed as

$$\dot y = \dot R\left(x - t\right)R^\dagger + R\left(\dot x - \dot t\right)R^\dagger + R\left(x - t\right)\dot R^\dagger$$

The first and last terms are reverses of each other and can be combined. The last term can be split into two terms to yield

$$\dot y = 2\left\langle \dot R\left(x - t\right)R^\dagger \right\rangle_1 - R\dot t R^\dagger + R\dot x R^\dagger \qquad (8.2.1)$$

This may be interpreted term by term starting from the last one.

The last term on the right is a function of the intrinsic velocity of the point of interest (e.g. a particle velocity), $\dot{x}$, after it has been rotated into an expression in the body frame.

The middle term on the right is an apparent point velocity (as observed in the body frame) which is due to motion of the body itself (e.g. motion of the origin of the body frame wrt the reference frame) as expressed in the reference frame, $\dot{t}$, after it is rotated to be expressed in the body frame.

The first term uses the notation

$$\left\langle \dot{R}\left(x-t\right)R^{\dagger}\right\rangle_{1} = \frac{1}{2}\left(\dot{R}\left(x-t\right)R^{\dagger} + R\left(x-t\right)\dot{R}^{\dagger}\right)$$

This represents the "vector part" of the entity inside the brackets (which here is a GA element comprising a vector and trivector part). This vector represents an apparent orbital velocity, that is observed in the body frame, due to changes in attitude of the body wrt the reference frame.

8.2.2. Point Velocity and Rotational Velocity

Substitution of relationship 8.1.2 into the point velocity relationship of equation 8.2.1, allows expressing the body frame velocity, $\dot{y}$, as

$$\dot{y} = \left\langle \Omega R\left(x-t\right)R^{\dagger}\right\rangle_{1} - R\dot{t}R^{\dagger} + R\dot{x}R^{\dagger}$$

In this expression, all entities on the right hand side are expressed in the reference frame coordinate system. This is often useful for associating physical interpretation and meaning to parameter values in the context of dynamical situations.

The last three factors in the first term, represent the point of interest expression in the body frame, y, (ref. relationship C.5.1), so that the body frame velocity may be expressed as an implicit, or "mixed-domain", expression,

$$\dot{y} = \left\langle \Omega y\right\rangle_{1} - R\dot{t}R^{\dagger} + R\dot{x}R^{\dagger} \tag{8.2.2}$$

This mixed-domain formulation is often useful when working with various sensors (e.g. such as when the point of interest at, y, may represent the location of a specific sensor element).

8.2.3. Point Velocity and Angular Velocity

For computational work, the Ω and R, quantities in above equation 8.2.2 may be expressed in terms of the bivector angle parameter, H, by substituting the relationships described in Sections 8.1.2 and 8.1.3 to obtain,

$$\dot{y} = 2\left\langle \overline{e^H e^{-H} y} \right\rangle_1 - e^H \dot{t} e^{-H} + e^H \dot{x} e^{-H}$$

An expression by which the spinor derivative, $\overline{e^H}$, may be evaluated in terms of half-angle, H and half-angle velocity, $\dot{H}$, is described subsequently in Section 10.1.3.

The main concept to note here, is that the apparent point velocity (as observed in the body frame) is expressed as a function of several quantities, e.g.

$$\dot{y} = \dot{y}\left(y, H, \dot{H}, \dot{t}, \dot{x} \right)$$

This is an implicit relationship involving the body frame position, y, However, this form is often useful in practice (such as when y represent the location of a strapdown sensor module and its position is a fixed and known in the body frame).

Alternatively y can be expressed in terms of the canonical transform 6.0.1, to obtain

$$\dot{y} = 2\left\langle \overline{e^H} \left(x - t \right) e^{-H} \right\rangle_1 - e^H \dot{t} e^{-H} + e^H \dot{x} e^{-H}$$

Now, the apparent point velocity expressed in the body frame, is a function of only the free parameters of state. I.e.

$$\dot{y} = \dot{y}\left(H, \dot{H}, \dot{t}, \dot{x} \right)$$

This is a non-linear, first order, differential equation that *represents the full kinematic state* of a physical body with a *minimum number of parameters* and is free of mathematical or physical singularities and is free from any implicit assumptions or additional constraints!!!

8.3. Point Acceleration in the Body Frame

The differentiation process above for velocities can be repeated to obtain an expression for acceleration. The GA operations are straight forward, although not particularly illuminating and are elided here. The result produces an acceleration relationship of which one form, useful for interpretation is,

$$\ddot{y} = \left\langle \dot{\Omega} y \right\rangle_1 + 2 \left\langle \Omega \dot{y} \right\rangle_1 + \left\langle \langle \Omega y \rangle_1 \Omega \right\rangle_1 + R \ddot{t} R^\dagger + R \ddot{x} R^\dagger \quad (8.3.1)$$

This is an expression for the acceleration of the point of interest as the acceleration is expressed in the body frame. I.e. this is the acceleration observed by an accelerometer sensor attached to the body!

The rotational acceleration, $\dot{\Omega}$, is connected to the angular velocity parameters as described in Section 8.1.3.

The first term on the right represents a observed (aka "fictitious") acceleration associated with a change in the bodies rotational velocity (e.g. such as due to torques on the body) and is sometimes called "Euler acceleration". The second term is the observed "Coriolis acceleration". The third term is the observed "centripetal acceleration". The fourth term is an observed acceleration due to acceleration of the body itself. The last term is associated with the actual (aka "proper") acceleration of the point of interest.

If an accelerometer is fixed to the body (e.g. a "strapdown" sensor), then it has a zero body-frame velocity and the second term on the right is zero. Another special case, although generally of less practical use, is if an idealized (infinitely small) accelerometer is coincident with the body coordinate frame origin ($y = 0$). In that case, all of the first three terms on the right vanish.

All terms express an acceleration that is observed in the body frame. In this particular, mixed-domain and implicit representation (i.e. body frame parameters, y and $\dot{y}$, also appear on the right side), the first three terms include parameters that are expressed in the body frame. E.g. the expression as written here represents an implicit function.

An explicit function expression is readily obtained simply by substituting the appropriate defining expressions for y and $\dot{y}$, after which the entire right hand side contains only parameters expressed with respect to the reference frame..

$$
\begin{aligned}
\ddot{y} = {}& \left\langle \dot{\Omega} R x R^\dagger \right\rangle_1 + \left\langle \left\langle \Omega R x R^\dagger \right\rangle_1 \Omega \right\rangle_1 + 2 \left\langle \Omega \left\langle \Omega R x R^\dagger \right\rangle_1 \right\rangle_1 \\
& - \left\langle \dot{\Omega} R t R^\dagger \right\rangle_1 - \left\langle \left\langle \Omega R t R^\dagger \right\rangle_1 \Omega \right\rangle_1 - 2 \left\langle \Omega \left\langle \Omega R t R^\dagger \right\rangle_1 \right\rangle_1 \\
& + 2 \left\langle \Omega R \dot{x} R^\dagger \right\rangle_1 \\
& - 2 \left\langle \Omega R \dot{t} R^\dagger \right\rangle_1 \\
& + R \ddot{x} R^\dagger \\
& + R \ddot{t} R^\dagger
\end{aligned}
\tag{8.3.2}
$$

For computational work, the Ω, $\dot{\Omega}$ and R entities above may be expressed and evaluated in terms of the bivector angle parameter, H and its derivatives. This result is an expression for acceleration (expressed in body frame) defined in terms of the precise 6 dof components associated with bivector, H, and vector, t, and their associated derivatives,

$$
\ddot{y} = \ddot{y}\left(H, \dot{H}, \ddot{H}, t, \dot{t}, \ddot{t} \right)
$$

Note that all arguments to this function, $\ddot{y}$, as with all terms on the right hand side above, are expressed in the reference frame.

This form represents an explicit function formulation of the body frame acceleration. As such, it is perhaps a bit more "elegant". It is particularly well suited for work with inertial measurement and navigation systems.

However, in many practical scenarios of interest, the implicit form above is often very useful because it may be more closely associated with data that are often available. The mixed-domain implicit representation is often particularly useful for instrumented bodies (e.g. aircraft with onboard navigation systems, devices with MEMS sensors such vehicles and smart phones, etc).

Overall, a main emphasis here is that the acceleration expressions offer relatively simple analytical expressions for quantities that often can be observed, directly and quantitatively. The explicit forms are often useful for observation *by external observers and systems* that track a rigid body (e.g. active motion tracking systems). The mixed-domain implicit formations are often useful for including observations made by sensors and systems that are *mounted on the moving rigid body* (e.g. inertial systems).

Part IV.

Practical Considerations

9. Observations and Parameter Recovery

It's fine to be able to represent rigid body states (static or kinematic) analytically. However, to have real value, these formulations need to be useful for accomplishing work in practice.

Practical processes often involve a combination of actions such as the following.

Modeling
> Create a mathematical formulation that represents the state(s) of a physically important rigid body(ies) of interested

Prediction
> Use the model to make predictions for "observable" quantities (e.g. things that can be "measured")

Acquisition
> Use corresponding physical system to capture and record measurements (e.g. images, inertial measurement time series, etc)

Fitting
> Use the acquired observational data to determine various elements of the body state (e.g. "solve for" body position, attitude, velocities, etc)

Utilization
> Use the "fit" parameter values to make predictions or secondary measurements needed for specific applications (e.g.

to formulate and issue command and control actions that affect body motion, or e.g. interpolate/extrapolate a sparse collection of measurements)

The basic rigid body "modeling" concepts are addressed in the previous parts and chapters. The "acquisition" process is primarily associated with actual physical systems while the "utilization" step is entirely application specific. This leaves the two stages of "prediction" and "fitting" to be addressed presently.

The scope and sophistication of the prediction and fitting stages depends on what is practical during the acquisition stage and what is needed for the utilization stage. To maintain a focus on the concepts involved and not be bogged down with application specifics, the following sections of this chapter will address a specific example case application - that of body state determination.

9.1. Example - State Determination

The determination of static state can be summarized as "given observations of various geometric entities, determine the state of a rigid body". To keep things simple for illustration, here it is assumed that the observations are in the form of "corresponding point coordinates".

I.e. a set of three or more point locations are measured in both the reference frame, the $\{x_n\}$, and in the body frame, the $\{y_n\}$. Given these two sets of correspondences, it is desired to determine the static body state defined as "best fit" values to parameters, t and H.

For this example, assume the reference frame point locations, x_n, are known and the corresponding body frame locations, y_n, are to be measured and the rigid body state parameters, t and H, are to be estimated.

In the following sections, an over-set tilde (e.g. $\tilde{\bullet}$) is used to denote a measured value (with implied known uncertainty properties) while a over-set check (e.g. $\check{\bullet}$) is used to indicate a "fit" or "solution" value for a given parameter.

9.1.1. Modeling

For this illustrative and simple state estimation example, the mathematical model is simply the canonical rigid body transformation relationship expressed by equation 6.0.1.

9.1.2. Prediction

In this simple (nearly trivial) example case, the prediction stage consists of predicting the values of y_n values given the measured value x_n with the prediction being a function of the body state parameters. The prediction relationship is the canonical rigid body transformation expression 6.0.1 rewritten slightly to denote the measured quantities for the n-th pair of corresponding measurements.

$$\tilde{y}_n - e^H \left(\tilde{x}_n - t \right) e^{-H} = v_{xyn}$$

where the vector, v_{xyn}, represents a residual "algebraic misclosure" that is associated with errors in the measurement values.

NOTE: This is a convenient formulation with which to describe a simple example. However, it is generally *NOT* the appropriate prediction model for this kind of problem. In practice, for a "real-world" problem like this, it would be desirable to minimize the error associated with each measurement. E.g. to formulate the prediction model in terms of direct observation residuals such as

$$\tilde{y}_n - \check{y} = v_{yn}$$
$$\tilde{x}_n - \check{x} = v_{xn}$$

Then purse a solution that minimizes the v_{yn} and v_{xn} vector residuals. However, this level of detail adds unnecessary complexity for present purpose of describing the high level concepts.

To further simplify the example for illustration, introduce abstract algebraic residual vectors, v_n, that have no particular physical significance. As just noted, this is typically a questionable approach in practice, but can be useful for "quick and dirty" solutions and/or as an approximation that is to be refined subsequently.

Express the canonical rigid body transform relationship 6.0.1 as

$$y - e^H \left(x - t \right) e^{-H} = 0$$

Multiply this from the right by e^H to obtain, the imaginary spinor relationship (an expression in which all the terms have vector and trivector grades)

$$y e^H - e^H x - e^H t = 0$$

Now introduce the abstract algebraic residual, v_n, to absorb errors in the measured values,

$$\tilde{y}_n e^H - e^H \tilde{x}_n - e^H t = v_n \qquad (9.1.1)$$

9.1.3. Acquisition

For this example, the acquisition stage is associated with making the required measurements. E.g. using electronic devices, retractable tapes, survey instruments, or some other form of coordinate measuring technology to obtain coordinates for the points of interest - both the coordinates expressed with respect to the reference frame, i.e. the $\tilde{x}_n$, and also the coordinates expressed with respect to the body frame, i.e. the $\tilde{y}_n$.

9.1.4. Fitting

The fitting step has the goal of determining model parameters (here the rigid body state parameters) that are consistent with the predicted observation relationships. I.e. for the present example case, determine values for (the 6 components of) t and H that are consistent with multiple observations of the form of equation 9.1.1.

There are many approaches and technique to determine values for the rigid body state parameters. One of the more elegant approaches is to manipulate the system of observation equations into a three-part problem that can be solved by simple algebraic operations and one matrix eigenvector decomposition (e.g. [6]). However, once again, specific solution approaches are not the point here. The current point is that there is a system of observations of the form of equations 9.1.1.

For mathematical rigor, introduce an observation function, f, defined as

$$f_n(t, H) = \tilde{y}_n - e^H \tilde{x}_n - e^H t = v_n \qquad (9.1.2)$$

The point of the data fitting stage is to determine "good" values for the arguments of function, f, that produce overall small magnitude values for the v_n.

For example, a classic formulation of "good" is to seek parameter values such that the sum of squared residuals is minimum. E.g. find parameter values, $\check{t}$ and $\check{H}$, that minimize the value of a merit function, χ^2

$$\chi^2 = \sum_n v_n^2$$

For the present (very simple) case, this merit function is the same as

$$\chi^2 = \sum_n f_n^2$$

This is a classic non-linear "least squares" problem and can be addressed with well-known relevant methods and techniques.

I.e. given the input values, $\tilde{x}_n$ and $\tilde{y}_n$, find parameter values, $\check{t}$, and $\check{H}$, that minimize χ^2.

For many of these methods, it is useful (or required) to be able to express derivatives of the observation functions. With the t and H parameter formulation of rigid body state, this is very easy. E.g. the full derivative of the function associated with observation 9.1.1 is

$$\dot{f}_n = \dot{y}_n - \overline{e^H}\left(x_n + t\right) - e^H\left(\dot{x}_n + \dot{t}\right)$$

The derivatives are described further in Chapter 10.

9.1.5. Utilization

Continuing with the simple illustrative state estimation example, the fitting process provides parameter values, $\check{t}$, and $\check{H}$, that describe (a best estimate of) the state of the rigid body. Some hypothetical application might then use this estimated state to perform various computations.

For example, the coordinates of other points expressed in the reference system could be transformed into an expression in the body frame. Or the transformation relationship could be inverted and used to compute reference frame coordinates for point locations that are known in the body frame.

10. Useful Differentiation Relationships

The following uses both and overdot and also an overline nota-tion as alternative symbols for denoting differentiation wrt an arbitrary scalar. The overline allows emphasizing and clarifying the scope of the entity to which the differentiation applies.

Returning to the idea of determining numeric (or best fit) values for the 6 parameters describing rigid body state - it's necessary to solve some form of non-linear system. Typically this involves some version of a "guess and refine" algorithm. For refinements, it is often advantageous to have derivatives (also very useful for sensitivity analysis).

The velocity relationship of equation 8.2.1 can be expressed in terms of the 3 vector and 3 bivector parameters as follows.

$$\dot{y} = 2\left\langle \overline{e^H}\,(x-t)\,e^{-H} \right\rangle_1 - e^H \dot{t} e^{-H} + e^H \dot{x} e^{-H} \qquad (10.0.1)$$

For interpretation and insight, the derivative of the exponen-tial factor may be expressed in terms of magnitude and direc-tion quantities (For numerical computation it's typically more efficient and stable to use formulae involving the three scalar components of bivector H directly as described further below).

$$\overline{e^H} = \overline{e^{\vartheta\Theta}} = e^{\vartheta\Theta}\dot{\vartheta}\Theta + \dot{\Theta}\sin\vartheta$$

In terms of the magnitude-direction decomposition of $H = |H|\,\hat{H}$,

$$\overline{e^H} = e^H \overline{|H|}\hat{H} + \overline{\hat{H}}\sin|H|$$

The first term represents change that is due to change in the magnitude of the rotation angle. The magnitude of the term is proportional to the rate of change of the rotation angle magnitude, and the direction is a quarter turn out of phase (the $\frac{\pi}{2}$ value).

The second term represents change that is due to changing the plane of rotation (e.g. such as often occurs in the presence of torques on free moving bodies). The magnitude of this term is proportional to both the magnitude of the change in direction of the plane, and to the sine of the angle through which the object is already rotated.

Aside from the useful physical interpretations, a key point here is that a simple analytical expression for the derivatives exists!

The above particular expression is typically not the best for practical computations. There is a more generally practical way to compute the exponential derivative. This is addressed in the following sections and also Appendix D which describe computationally useful expressions.

10.1. Gradients and Curvatures

10.1.1. Gradients / Velocities

Expressing the derivative relationships in terms of the underlying vector and bivector parameters is quite simple. For example, in terms of parameters, t and H, the rigid body "velocity" transformation relationship 10.0.1 includes the set of parameters derivatives,

$$\left\{ \dot{t}, \dot{H}, \overline{e^H} \right\}$$

The spinor derivative (rotational velocity), $\overline{e^H}$, can be expanded in terms of H and $\dot{H}$ with the expressions of Section 10.1.3 below. This allows the rigid body transformation velocities and

61

gradients to be expressed in terms of the set of parameter values,

$$\left\{t, \dot{t}, H, \dot{H}\right\}$$

These parameters exactly cover the expected physical 12 dofs (6 for state position and 6 for state velocities).

10.1.2. Curvatures / Accelerations

In corresponding fashion, derivatives of the "acceleration" transformation relationship 8.3.2 include the set of parameters,

$$\left\{t, \dot{t}, \ddot{t}, R, \Omega, \dot{\Omega}\right\}$$

The spinor, R, can be represented in terms of H (per defining relationship 8.1.1, so acceleration relationships may be expressed in terms of rotational velocities and rotational accelerations with the parameter set,

$$\left\{t, \dot{t}, \ddot{t}, H, \Omega, \dot{\Omega}\right\}$$

The rotational velocity, Ω, and the rotational acceleration, $\dot{\Omega}$, may be expanded in terms of the spinor exponential function, e^H, and its first and second derivatives as in equations 8.1.4 and 8.1.5 and then expressing or evaluating first and second order derivatives as described in the following Sections 10.1.3 and 10.1.4. This allows specifying the parameter curvature relationship in terms of the parameter set

$$\left\{t, \dot{t}, \ddot{t}, H, \dot{H}, \ddot{H}\right\}$$

This collection of parameters is associated with angular velocity and angular accelerations and comprises 18 dof as expected (6 for state position, 6 for state velocity and 6 for state acceleration).

10.1.3. Spinor First Derivative

The term, $\overline{e^H} = \dot{R}$, is the rotational velocity (not the angular velocity). It may be evaluated easily in terms of the angle, H, and angular velocity, $\dot{H}$, using the appendix relationship D.1.1. I.e. computing the exponential derivative via

$$\overline{e^H} = \begin{cases} \frac{1}{2}H^{-2}\left(H\dot{H} + \dot{H}H\right)He^H & \\ -\frac{1}{2}H^{-2}\left(H\dot{H} - \dot{H}H\right)\frac{1}{2}\left(e^H - e^{-H}\right) & \varepsilon < |H| \\ \dot{H} + \left(H \cdot \dot{H}\right)\left(1 + \frac{1}{2}H\right) + \mathcal{O}\left(H^3\right) & |H| \le \varepsilon \end{cases}$$

$$(10.1.1)$$

10.1.4. Spinor Second Derivative

Computing the second derivative of the spinor, e^H, is simply a matter of differentiating the function relationships in equation 10.1.1. This is not difficult, but is a bit lengthy.

To simplify expressions, first express $\overline{e^H}$ from equation 10.1.1 in terms of intermediate variables,

$$\overline{e^H} = \lambda\mu He^H - MN$$

Here the intermediate quantities are: scalars, λ and μ, along with bivectors, M and N, defined as

$$\lambda = \frac{1}{2}H^{-2}$$
$$\mu = H\dot{H} + \dot{H}H$$
$$M = H\dot{H} - \dot{H}H$$
$$N = \frac{1}{2}H^{-2}\frac{1}{2}\left(e^H - e^{-H}\right) = \frac{1}{2}|H|^{-3}H\sin|H|$$

The equivalence on the last line arises from the definition of the sine function which is defined as the sum including only the "odd" terms (the bivector-valued grades) from the exponential series expansion.

Differentiate this by applying the product rule to each term to obtain

$$\overline{\dot{e^H}} = \dot{\lambda}\mu H e^H + \lambda\dot{\mu}H e^H + \lambda\mu\dot{H}e^H + \lambda\mu H\overline{\dot{e^H}} - \dot{M}N - M\dot{N}$$

The intermediate quantity derivatives may be expressed (after a touch of basic algebra and calculus) as

$$\dot{\lambda} = -|H|^{-2}\lambda\mu \tag{10.1.2}$$

$$\dot{\mu} = 2\dot{H}^2 + \left(H\ddot{H} + \ddot{H}H\right) \tag{10.1.3}$$

$$\dot{M} = \left(H\ddot{H} - \ddot{H}H\right) \tag{10.1.4}$$

$$\dot{N} = \frac{1}{2}|H|^{-3}\left(\left(3\lambda\mu H + \dot{H}\right)\sin|H| - H\cos|H|\right) \tag{10.1.5}$$

The spinor second derivative is expressed in terms of the bivector quantities of: the 3D angle, the 3D angular velocity, and the 3D angular acceleration. I.e. together, the above groups of relationships provide an expression for the second derivative, $\ddot{\exp}(H)$, in terms of free angle parameter and its first and second derivatives

$$\left\{H,\ \dot{H},\ \ddot{H}\right\}$$

For Small Angles

To address small angles, $|H| < \varepsilon$, first replace equation 10.1.1 with one of the small angle representations described in Appendix D.2 (e.g. equation D.2.2 or D.2.3), then differentiate that spinor velocity expression. E.g. using the third order series expansion,

$$\overline{e^H} = \dot{H} + \left(H \cdot \dot{H}\right)\left(1 + \frac{1}{2}H\right) + \mathcal{O}\left(H^3\right)$$

After applying the differentiation product rule to each term and performing some simple algebra, this may be expressed as

$$\overline{e^H} = \ddot{H} + \dot{H}^2 + \left(\ddot{H} \cdot H + \frac{1}{2}\dot{H}^2 H + \frac{1}{2}\dot{H}\left(\dot{H} \cdot H\right)\right) + \mathcal{O}\left(H^2\right)$$

10.1.5. Parameter Derivatives

The parameters derivatives themselves, $\dot{H}$, and $\dot{t}$, depend on the type of coordinate system in use. However, when using fixed-axis Cartesian reference frames, these parameter derivatives are exceptionally simple.

Consider parameters expressed in component form as

$$t = \sum \tau_k e_k$$
$$H = \sum \eta_k E_{ij}$$

where it's understood that the summations run $k = 1, 2, 3$, and the combined indices occur only in cyclic order.

For this Cartesian expression, the derivatives with respect to individual parameter components[1] are given directly by the coordinate frame basis elements. I.e.

$$\frac{\partial t}{\partial \tau_k} = e_k$$

---

[1] Derivatives with respect to the parameters components are used, for example, in Jacobian and Hessian matrices involved with numeric solution methods.

10. Useful Differentiation Relationships

$$\frac{\partial H}{\partial \eta_k} = E_{ij}$$

For kinematic functions as components, the derivatives wrt an arbitrary scalar sequencing function are as expected,

$$\dot{t} = \sum_k \dot{\tau}_k\left(\sigma\right)\dot{\sigma}$$

$$\dot{H} = \sum_k \dot{\eta}_k\left(\sigma\right)\dot{\sigma}$$

And, for the special case, where the scalar *is* a fundamental parametric sequence index, σ, the derivatives simplify to

$$\frac{\partial t_k}{\partial \sigma} = \dot{\tau}_k\left(\sigma\right)$$

$$\frac{\partial H}{\partial \sigma} = \dot{\eta}_k\left(\sigma\right)$$

Part V.

Concluding Remarks

11. Practical Observations

It should be emphasized that for *everything* presented herein, there are 6 and only 6 free parameters in the math models[1]. *The 6 free math parameters exactly match the 6 degrees of freedom in the physical configuration.*

The presented rigid body math model contains *no unnecessary* degrees of freedom and therefore is *free of unnecessary algebraic additions or supplements*[2]. There is no need to consider, represent, nor incorporate algebraic constraint concepts or any implied relationships.

There is significant practical value in the 6 math parameters being associated directly with the 6 physical dof of a rigid object. The association is often "one-to-one" with important qualities and relationships that *can be observed directly* using physical instrumentation and sensors!!

In addition, this formulation provides relatively simple *analytical expressions for all orders of derivatives*. Computationally useful first and second order derivative expressions are presented herein. These expressions are sufficiently simple and compact for easy computation and implementation in computer programs.

Overall, the author has *used this type of modeling and method-*

---

[1] In this context, spinors can be consider to be an algebraic convenience. Although they may be expressed in Cartesian coordinates in terms of four components and a constraint between them, at any time, the spinor can be replaced by its association with the exponential of the 3 dof bivector angle parameter.

[2] Such as additional independent constraint equations or implicit order-of-operations conventions.

ology with great success and over a wide range of project types and application scenarios across many industries and disciplines. From those experiences, it is easy to say, and to say without qualification, that this approach "works *really* well".

Part VI.

Appendices

A. Useful Geometric Algebra Items

A.1. General Entities

In three-dimensional (3D) space, the fundamental constructs of geometric algebra are the entities known as: scalars, vectors, bivectors, and trivectors. Each of these fundamental types is known as a "blade" and represents a geometrically meaningful algebraic subspace.

Scalars are 0-vectors and represent "amounts". Vectors are 1-vectors and represent directed linear segments. Bivectors are 2-vectors and represented directed planar segments. Trivectors are 3-vectors and represented directed volume segments.

A.1.1. Scalars

Scalar values represent oriented (i.e. $\pm$) quantities (0D subspaces) and are denoted herein with Greek letters. Scalars as used here are associated with Real numbers. I.e. scalars *are* members of the algebra of real numbers, $\mathfrak{R}$.

The operation of scalar multiplication changes the size (e.g. "scale") of any algebraic element being multiplied. Multiplication by a positive scalar value preserves the direction of the other factor, while multiplication by a negative scalar value changes the direction of the other factor.

Scalar value multiplication is commutative with all members of the algebra. E.g. specifically useful herein is that scalar multiplication is commutative with vectors and bivectors.

A.1.2. Vectors

Vectors represent oriented lines (1D subspaces) and are denoted herein with lowercase Latin letters. In physical terms, they represent a from-to relationship between two individual points.

Interpreted with the conventions employed herein, vectors represent the position of an "arriving-at" point relative to a "starting-from" point (e.g. interpreting a vector as "tip wrt tail".

When considered in a mathematical sense, the vectors used here are exactly those of defined in classic linear algebra (although only 3D spaces are needed herein). When considered physically, the vectors here are identical with those used in engineering and classical mechanics (tracing their roots to Gibbs' interpretation of Hamilton's quaternions).

In 3D space, the expression of a vector is associated with three free parameters. When expressed in coordinate systems with fixed, orthogonal direction axis (i.e. Cartesian reference frames).

The 3D Cartesian frames used herein, can be associated with three basis vectors, e_k, for $k = 1, 2, 3$. The e_k are a set of orthonormal basis vectors that define the Cartesian coordinate frame's alignment reference (e.g. the e_1, e_2 and e_3 are the "x", "y", and "z" axis directions). Each e_k is unit magnitude such that $e_k^2 = 1$. The collection of e_k are all mutually orthogonal and e_i, e_j, and e_k, for i, j, k in cyclic order, form a dextral (right-handed) triad ($e_i e_j e_k$ is a dextral unitary trivector).

In Cartesian reference systems, vectors, such as an arbitrary vector, a, have a particularly simple component expression (e.g. a vector's Fourier decomposition into orthogonal constituents),

$$a = \sum_k \alpha_k e_k = \alpha_1 e_1 + \alpha_2 e_2 + \alpha_3 e_3$$

Each of the α_k is a scalar and indicates "how much" of this vector projects (orthogonality) onto the k-th coordinate frame axis.

In 3D space, specifying values for each of the three independent coefficients, α_k, completely and uniquely defines the vector, a.

All non-zero vectors can be factored into the product of a scalar magnitude and a vector direction,

$$a = |a|\,\hat{a}$$

where

$$|a| \equiv \sqrt{a^2} = \sqrt{\sum_k \alpha_k^2}$$

For $a \neq 0$,

$$\hat{a} = \frac{1}{|a|}a$$

For $a = 0$, the direction is not uniquely defined.

It is worth noting that the vector, a, remains completely well-defined even for $a = 0$. It is only this particular magnitude and direction factorization that lacks uniqueness[1].

A.1.3. Bivectors

Bivectors represent oriented planes (2D subspaces) and are denoted herein by capital Latin (e.g. B) or capital Greek letters (e.g. Θ).

With respect to a Cartesian coordinate system, a bivector can be expressed as a sum of three linearly independent components. E.g.

$$B = \beta_1 E_{23} + \beta_2 E_{31} + \beta_3 E_{12} \qquad (A.1.1)$$

---

[1] E.g. in the limit as $a \to 0$, the direction expression tends to, $\frac{0}{0}$, indicating that it is not appropriate to attempt this particular factorization for $a = 0$.

A. Useful Geometric Algebra Items

Here, the β_k are scalar "components" (just like the α_k are scalar components of vector described above) and the E_{ij} are an orthonormal set of basis bivector planes (just like the e_k are an orthonormal set of basis vector axes).

The E_{ij} unit bivectors describe the "basis bivectors" of the coordinate frame (e.g. the positively oriented cardinal planes), analogous to how the e_k unit vectors describe the basis axis of the coordinate frame (e.g. the positively oriented cardinal axis directions). Specifically, E_{23}, E_{31}, E_{12} can be associated with the "YZ", the "ZX" and the "XY" planes[2].

It's worth mentioning that the E_{ij} basis bivectors can be defined in terms of the e_k basis vectors, specifically as $E_{ij} \equiv e_i e_j$. However, this is not particularly important for present purposes.

A particularly interesting and profound feature of bivectors, including the basis bivectors, is that they square to non-positive scalar values (exactly like the "imaginary" basis element of Complex numbers). In particular,

$$E_{ij}^2 = -1$$

This can be used to show that any bivector, B, also squares to a non-positive value,

$$B^2 \leq 0$$

In general, bivectors, just like vectors, can be decomposed into scalar-valued magnitude and vector-valued direction factors, just as vectors can be. E.g.

$$B = |B|\, \hat{B}$$

---

[2]The ordering here is a convention which maintains the cyclic index interpretation, e.g. $B = \sum_{\text{cyclic}} E_{ij}\beta_k$, which is useful for keeping algebraic signs straight (and also for implicit summation notations). This notation is consistent with the unit positive trivector, $\mathcal{I}$, being expressible as $\mathcal{I} = e_i E_{jk} = e_i e_j e_k$ for indices, i, j, k taken in cyclic order.

where

$$|B| = \sqrt{-B^2} = \sqrt{\sum_k \beta_k^2}$$

For bivectors, the magnitude definition involves an algebraic negation because bivectors square to non-positive scalar values.[3],

The unitary direction factor for $B \neq 0$, is defined in the same way as for vectors, i.e.

$$\hat{B} = \frac{1}{|B|}B$$

For $B = 0$, no direction is defined. Again, as with vectors, $B = 0$, is very well defined and well behaved, and it is only the magnitude/direction factorization that is undefined for zero bivectors.

A.1.4. Trivectors

Trivector entities complete the GA blades spanning 3D space. Akin to how vectors represent directed lines, and how bivectors represent directed planes, trivectors represent oriented volumes. Positive direction trivectors represent dextral (right-handed) volumes while negative trivectors represent sinister (left-handed) volumes.

Trivectors are an integral aspect of the overall algebra, but are not directly needed for purposes herein. However, it should be mentioned that, in 3D space, trivectors act like "imaginary scalars".

---

[3]Actually, the magnitude is defined in more general terms as the square root of the product of the bivector and it's reverse, i.e. $|B| = \sqrt{BB^\dagger}$. However, since $B^\dagger = -B$, the end result is introduction of a simple negative sign under the radical. Another way to consider this, is that negative values result from the various E_{ij}^2 terms that occur when the Cartesian expansion of $B = \sum \beta_i E_{jk}$ when they are multiplied by themselves, in which case the leading minus sign is required to counter this.

In general, trivectors, like bivectors, square to non-positive values and therefore provide another form of "imaginary" number.

Whereas bivectors in general provide a kind of "*directed* imaginary number", trivectors in 3D space provide an "*undirected* imaginary number"[4].

In 3D space, trivector multiplication commutes with all other entities in the algebra.

A.1.5. Multivectors

Multivector entities are the composition of a scalar, a vector, a bivector and a trivector, any of which may be present or absent (in which case the missing grade is interpreted s having a zero value). E.g.

$$\mathrm{M} = \mu + m + M + \mathcal{M}$$

Angle brackets are used to indicate the grade of the element. I.e.

$$\mathrm{M} = \langle M \rangle_0 + \langle M \rangle_1 + \langle M \rangle_2 + \langle M \rangle_3$$

The grades are associated with constituents of the multivector above as,

$$\langle \mathrm{M} \rangle_0 = \mu$$
$$\langle \mathrm{M} \rangle_1 = m$$
$$\langle \mathrm{M} \rangle_2 = M$$
$$\langle \mathrm{M} \rangle_3 = \mathcal{M}$$

The reverse operation applied to a multivector changes the signs of the bivector and trivector grades and may be expressed as,

$$\mathrm{M}^\dagger = \mu + m - M - \mathcal{M}$$

---

[4]In space dimensions higher than 3D, the volumes associated with trivectors are also "directed" within the larger dimensional space. In 2D space, no trivectors exist.

A.1.6. Spinors

Spinors are of fundamental importance in representing rotations because they appear naturally in the canonical rotation formula of the form (cf expressions 6.0.1 and 8.1.1).

Spinors are a special case of multivectors that comprise only two grade constituents, a scalar and a bivector grade. E.g. A spinor R, is a multivector with zero vector and zero trivector parts, as

$$R = \langle R \rangle_0 + \langle R \rangle_2$$

and it has reverse,

$$R^\dagger = \langle R \rangle_0 - \langle R \rangle_2$$

Spinors may be represented as the exponential of another spinor. E.g.

$$R = e^{\langle Q \rangle_0 + \langle Q \rangle_2}$$

If the scalar grade of the exponent argument is zero (e.g. Q is a pure bivector such that $\langle Q \rangle_0 = 0$ and $Q = \langle Q \rangle_2$), then the spinor, R, is unitary, i.e.

$$[\langle Q \rangle_0 = 0] \implies [|R| = 1]$$

This is particularly useful for expressing rigid body attitudes with bivectors. E.g. the half-angle, H, is a pure bivector, $H = \langle H \rangle_2$, so that for,

$$R_H \equiv e^H$$

then (on account of missing scalar grade in the exponent's argument)

$$|R_H| = 1$$

This leads to a generalization of Euler's formula (cf. [11]) which is actually just grade decomposition of the spinor, i.e.

$$e^H = R = \langle R \rangle_0 + \langle R \rangle_2$$
$$= \cos |H| + \hat{H} \sin |H|$$

Here the cosine term is a scalar, and the second term is scalar multiple (the $\sin|H|$) of the bivector direction (the $\hat{H}$),

Overall, this spinor, R_H, can be expressed in terms of four scalar components. When defined in terms of a pure bivector argument to the exponential function, the component values of R_H are constrained to sum-square to unity. The, R_H, defined here is isomorphic with the unitary quaternions that are often used to express rigid body attitude.

A key benefit here, is that this spinor expression can be reduced to the fundamental 3 dof whenever desired via the inverse function[5]

$$H = \ln{(R)}$$

and the result, H, has a fundamental and very useful physical interpretation.

In practice, it is often convenient to perform algebraic operations and develop mathematical models in terms of spinor formulations, and then, convert the resulting descriptive expressions into use of the angle formations for actual computation purposes. $\hat{H}$.

A.2. Vector-Vector Product

The product of one vector with another is a key (perhaps THE key) concept in geometric algebra. It is based on a methodology discovered by Hermann Grassmann with those techniques refined by William Clifford.

By introducing a *completely well-defined* vector-vector product one *that can be inverted algebraically*, geometric algebra provides an astonishingly powerful means of expressing geometric

---

[5]Because the forward function, e^H, is periodic there can be an ambiguity in the inverse operation. In general, the logarithm function $\ln{(e^H)}$, is interpreted as the principal branch angle, H, with size in the half-open range $[-\frac{\pi}{2}, \frac{\pi}{2})$.

relationships between objects in space and operating with those relationships algebraically.

The product of two vectors is NOT another vector, but is an entirely new entity. The new entity is a composite entity that comprises a scalar and bivector part as illustrated in the following.

A.2.1. In Cartesian Coordinate Form

The product of two vectors is a spinor comprising *both* a scalar and a bivector grade constituent.

In Cartesian coordinates the components of the spinor result can be expressed in terms of the components with which the two vectors are described. Assume two arbitrary vectors, a and c, expressed as,

$$a = \sum_k \alpha_k e_k$$

$$c = \sum_k \gamma_k e_k$$

Multiply these together (after introducing two unique "dummy" indices, i and j, for tracking the combinations of individual terms within the product),

$$ac = \left(\sum_i \alpha_i e_i \right) \left(\sum_j \beta_j e_j \right)$$

$$= \sum_{i,j} \alpha_i \beta_j \left(e_i e_j \right)$$

$$= \sum_{i<j} \alpha_i \beta_j \left(e_i e_j \right) + \sum_{i=j} \alpha_i \beta_j \left(e_i e_j \right) + \sum_{i>j} \alpha_i \beta_j \left(e_i e_j \right)$$

Note that:

A. Useful Geometric Algebra Items

- The product of orthogonal vectors, e_i and e_j, for $i < j$, defines the bivector basis entity, $E_{ij} \equiv e_i e_j$;

- The product of pair of orthogonal vectors is anti-commutative such that, $e_i e_j = -e_j e_i$; and

- The product of each unit vector with itself (i.e. its "square") is the unit scalar, $e_i^2 = 1$,

These considerations allow expressing the product

$$ac = \sum_{i=j} \alpha_i \beta_j + \sum_{i<j} \alpha_i \beta_j e_i e_j + \sum_{j<i} \alpha_i \beta_j \left(-e_i e_j \right)$$

Swapping the roles of the two dummy indices in the last term, then combining with the second term, leads to

$$ac = \sum_{i=j} \left(\alpha_i \beta_j \right) + \sum_{i<j} \left(\alpha_i \beta_j - \alpha_j \beta_i \right) E_{ij}$$

The first term on the right is the scalar grade constituent of the geometric product. The second term is the bivector grade.

Overall, the geometric product is often expressed in terms of the "dot" and "wedge" notation[6], which, for Cartesian coordinate frames, may be expressed as

$$ac = a \cdot c + a \wedge c$$

$$a \cdot c = \frac{1}{2} \left(ac + ca \right) = \sum_{i=j} \left(\alpha_i \beta_j \right)$$

$$a \wedge c = \frac{1}{2} \left(ac - ca \right) = \sum_{i<j} \left(\alpha_i \beta_j - \alpha_j \beta_i \right) E_{ij}$$

---

[6]The commonly used concept of 3D vector "cross" product can be expressed (much more correctly, with much more generality) in terms of the dual wedge product - i.e. $a \times b \equiv -\mathcal{I} \left(a \wedge b \right)$. As it turns out, this definition creates a true vector and completely avoids the needless kerfuffle of distinguishing "polar" and "axial" vectors as is required with the Gibbs' style vector cross product.

A. Useful Geometric Algebra Items

A.3. Bivector-Bivector Product

The product of two bivectors is a compound entity with the result comprising both scalar and bivector grade constituents.

In 3D space, the bivector-bivector product mirrors the vector-vector product[7].

The product of two bivectors is NOT another bivector. Rather it is an entirely new entity, a spinor, which is composed of two different grade elements.

A.3.1. In Cartesian Coordinate Form

For two bivectors, A, and B, expressed in terms of Cartesian components,

$$A = \alpha_1 E_{23} + \alpha_2 E_{31} + \alpha_3 E_{12}$$
$$B = \beta_1 E_{23} + \beta_2 E_{31} + \beta_3 E_{12}$$

the product may be expressed as

$$AB = - (\alpha_1\beta_1 + \alpha_2\beta_2 + \alpha_3\beta_3)$$
$$+ (\alpha_2\beta_3 - \alpha_3\beta_2) E_{23}$$
$$+ (\alpha_3\beta_1 - \alpha_1\beta_3) E_{31}$$
$$+ (\alpha_1\beta_2 - \alpha_2\beta_1) E_{12}$$

The term on the first line is a scalar (a grade 0 element), while the last three lines represent a bivector (a grade 2 element). This distinction can be very useful in practice. E.g.

$$AB = \langle AB \rangle_0 + \langle AB \rangle_2$$

---

[7]In more explicit terms, within 3D spaces, the bivectors are "dual" to vectors. E.g. for vector, b, a bivector may be defined as $B = \mathcal{I}b$, where $\mathcal{I}$ is the unit trivector, which is multiplicatively commutative. Thus, the product of two bivectors, A and C, can be expressed in terms of their dual vectors, a and c, as $AB = \mathcal{I}a\mathcal{I}b = \mathcal{I}^2 ab = -ab$. I.e. the bivector-bivector product is the negative of a related vector-vector product.

which may also be associated with a "dot" and a "commutator" notation[8],

$$AB = A \cdot B + A \otimes B$$

A key observation is that when a bivector is multiplied by itself, the bivector grade (commutator product portion) vanishes identically, leaving only the scalar grade portion

$$A^2 = -\left(\alpha_1^2 + \alpha_2^2 + \alpha_3^2\right)$$

This demonstrates that bivectors generally "square to negative values!". To be more precise, "the square of any bivector is a non-positive valued scalar".

A.4. Exponential Function

The importance of the exponential function in real algebra and calculus is well known. It is equally important in geometric algebra for rigid body descriptions and associated differential relationships.

The exponential function of an arbitrary quantity, Q, may be described, exactly as expected, in terms of the (absolutely convergent) algebraic series.

$$\exp(Q) = e^Q = \sum_{n=0}^{\infty} \frac{1}{n} Q^n$$

Here Q can be any multivector member of the geometric algebra of 3D space including the blades of the algebra (scalar, vector, bivector, trivector) as well as any general multivector (i.e. Q expressed as any combination of blades).

---

[8]This simple relationship is specific to 3D space. In higher dimensional spaces, the full decomposition for the bivector-bivector product is $AB = A \cdot B + A \otimes B + A \wedge B$. However, in 3D space, the "wedge" outer product vanishes identically, $A \wedge B = 0$.

A. Useful Geometric Algebra Items

The exponential function is well defined for *anything* that can be represented by an expression in GA. For present purposes, most interest lies with the case of Q being a bivector entity for which $Q = \langle Q \rangle_2$.

A.4.1. Exponential of a Bivector - a Spinor

For the particular special case of interest herein, the exponential argument is a bivector and the exponential can be computed readily.

The exponential, e^H, may be evaluated computationally using the standard sine and cosine functions for real numbers along with the angle direction, $\hat{H}$. For smaller angles, where this direction is not well defined numerically, the series expansion can be evaluated directly. Overall, the exponential of a bivector is easily evaluated as

$$e^H = \begin{cases} \cos|H| + \hat{H}\sin|H| & \varepsilon \leq |H| \\ 1 + H + \frac{1}{2}H^2 + \ldots & |H| < \varepsilon \end{cases}$$

Since the second line represents the first few terms of an alternating series, the magnitude of the truncation error is less than the magnitude of the first neglected term.

B. Bivectors and Rotation

B.1. Bivector as 3D Angle

A "3D angle, representing any finite rotation in 3D space, can be represented as a bivector. The concept of a bivector in GA represents a directed plane. In one useful form, a bivector can be expressed in terms of two vectors as follows[1].

Consider two arbitrary vectors, b and c. These define a unique bivector, B, which has properties including these:

1. B overall specifies the (3D bivector) angle through which a vector, b, should be rotated in order to come into alignment with the second vector, c. I.e. B specifies "the (3D) angle of c wrt b".

2. B has a direction, $\hat{B}$, that is interpreted to be the geometric plane containing both of the vectors, b and c.

   a) If b and c are (anti)parallel, then $B = 0$, and no unique plane is defined.

3. B has a "sign" that represents the direction (aka handedness) of the plane

---

[1] A seminal example of this is the definition of the E_{ij} basis as the product of the two orthogonal vectors, e_i and e_j, via $E_{ij} = e_i e_j$. For two arbitrary vectors, b and c, the combination, $\frac{1}{2}(bc - cb)$ is a pure bivector, B, (denoted by $b \wedge c$). E.g. $B = b \wedge c$.

B. Bivectors and Rotation

    a) The positive bivector, $+B$, defines a "forward face" of the geometric plane[2].

    b) The negated bivector, $-B$, represents the same geometric planar subspace but with opposite direction (e.g. corresponding to a "back face" of the plane).

4. B has a magnitude, $|B|$, which may be expressed in one particular form as

$$|B| = |c|\, |b|\, |\sin \vartheta|$$
                                            (B.1.1)

Here, ϑ is the size of the conventional size angle "between" the vectors. Specifically, it specifies the (signed) amount of turning motion necessary to rotate vector b into alignment with vector c.

    a) Note that (anti)parallel vectors define a zero bivector, $[c \parallel \pm b] \implies [B = 0]$.

    b) Note that bivector, B, is also zero if either vector b or c is zero.

    c) If b and c are both unit vectors (e.g. specifying only directions), then

$$0 \le |B| \le 1$$

or, alternatively,

$$-1 \le B^2 \le 0$$

---

[2] It is interesting and reassuring to observe that this direction (aka handedness) is *an intrinsic property of the plane* (an algebraic sign on the bivector). This means the direction of the plane can be defined within the plane itself and does *not* require any arbitrary external entity or construction, such as trying to define the direction by having a "normal vector" which may not even exist (e.g. does not exist in 2D space) or may be ambiguous (e.g. in 4D and higher dimension spaces).

As expressed above in item 4, expression B.1.1 is a bit of a simplification that is useful in describing the currently relevant points of the relationship[3] However, the main point of characteristic 4, is that two unit vectors[4], such as $\hat{b}$ and $\hat{c}$, can be used to express a bivector, and that the bivector completely and entirely represents a general angle in 3D space - not only the 1 degree of freedom associated with the size of the "classic" angle, but also the 2 additional dof associated with the geometric plane in which that angle size exists!!!

B.2. Bivector Physical Interpretation

To explain bivectors in physical terms, consider rotation of some physical object.

A completely general rotation can be accomplished by first deciding in what (arbitrary) direction to rotate. This defines the plane of rotation - e.g. a kind of "equator" for the rotation. Then rotate the object through some arbitrary magnitude angle (a scalar value in radians) while keeping the direction of the "equator" fixed in space[5].

Any general rotation can be expressed in this way. The final attitude of the object can be expressed in terms of a scalar size (1 dof) and a planar direction (2 dof). These three freedoms can be encoded into the three independent numbers that specify a general bivector. This is akin to the "angle/axis" and "Rodrigues Vector" formulations of rotation but the bivector representation is more algebraically consistent and powerful[6].

---

[3]In some ways, this bullet item is actually more of a "complification" since the relevant GA expression is actually much simpler (i.e. $B = b \wedge c$) - but that involves diving into a few additional algebraic interpretations not presently needed.

[4]Using the "hat" symbol to denote unitary magnitude quantities.

[5]Ref. Euler's theorem of rigid body rotation - cf. [9].

[6]With bivectors having a more consistent interaction and integration with

B.2.1. Angle Size and Direction

First consider specifying rotations of less than a half turn (for which the magnitude of the rotation angle is less than $|\pi|$). The direction of the rotation can be specified by the direction of an equatorial plane, which has a specific $\pm$ direction to it. Any final attitude can be reached by a rotation of magnitude, ϑ, with $0 \leq \vartheta \leq \pi$, that is performed with equatorial plane in either the $+\hat{B}$ direction, or else in the $-\hat{B}$ direction.

However, the sign on the direction of rotation, $\hat{B}$, can equally well be associated with the scalar part of the angle instead. E.g. by noting that,

$$\vartheta\left(\pm\hat{B}\right) = (\pm\vartheta)\,\hat{B}$$

Just as a bivector can be decomposed into a (non-negative) magnitude and a direction, it can also be decomposed into a (plus or minus) size and a direction.

Introduce the scalar, ϑ, to be interpreted as an angle "size" value that can be either positive or negative, so that

$$B = \vartheta\hat{B}$$

With this interpretation, the bivector, B, now encodes an angle that has a (uniquely oriented) plane of rotation, $\hat{B}$, in which the body can either rotate "forward" $(0 < \vartheta)$ or can rotate backward $(\vartheta < 0)$.

With this convention, any physically realizable body attitude can be expressed by a direction of the plane of rotation, $\hat{B}$, and a scalar angle size, ϑ, with $-\pi \leq \vartheta < \pi$.

If $B = 0$ (e.g. no rotation), the direction, $\hat{B}$, is not uniquely defined. However, the size of the angle is zero $(\vartheta = 0)$ so that the direction becomes irrelevant[7]

---

entire rest of the algebra.

[7] For zero rotation, the product representation, $\vartheta\hat{B}$, becomes a kind of

The above describes attitude of a body as it it situated in a static scenario. An angle size in the restricted "principal" range, $-\pi \leq \vartheta < \pi$, describes any possible static attitude and there is no reason to consider larger angle size ranges.

However, for kinematic situations when motion is involved, this is an unnecessary restriction. In general, the angle size, ϑ, can be unbounded, and larger sizes, $\pi < |\vartheta|$, correspond with multiple full rotations (e.g. can include any number of "full turns" that were completed "before" the body arrives at it's "current" attitude.

The bivector, B, supports encoding full cycles completely (e.g. the individual components, β_k, can get arbitrarily large and be either positive or negative).

B.2.2. Two Half-Rotations

Now consider (for reasons of mathematical convenience) that this rotation can be expressed in two separate, but equal and identical "half-rotations". E.g. define an (equatorial) plane of the rotation, then specify the half angle size, $\frac{1}{2}\vartheta$. Together these define a half-rotation. To obtain the final attitude of the rigid body, first rotate it by the half-rotation, and then rotate it again by the exact same half rotation once more.

Put another way, express the final attitude of the body by hypothetically performing an operation described as "a half-rotation operation that is applied twice in a row". I.e. the full rotation operation, $\mathcal{R}$, can be expressed in terms of a general half-rotation operator, $\mathcal{R}_H$ as

$$\mathcal{R} = \mathcal{R}_H\left(\mathcal{R}_H\right) = \mathcal{R}_H^2$$

$\frac{0}{0}$ quantity that has properties defined in the sense of limit processes. However, in practice, this is a non-issue, since practical computations are typically done with the three, β_k, components of B, those are always well defined for all rotations including no rotation at all (for which $B = 0$).

B. Bivectors and Rotation

In terms of rotation operations, this is like "applying the half-rotation" once, and then "applying the half-rotation" operation again to the result of the first half-rotation - i.e. "squaring[8] the half-rotation operation"[9].

This "two-times-half" approach accommodates the mathematical structure of the "sandwich style" multiplication of factors in the canonical rotation expressions such as equation C.4.1.

Returning to the physical interpretation, consider a unit magnitude direction vector, $\hat{b}$, that expresses the vector direction to some interesting feature on an object before it has been rotated. Then consider the unit vector, $\hat{c}$, that points to the location of that same interesting feature after the object has been *HALF-rotated*. This half-rotation can be interpreted as the direction, $\hat{b}$, "becoming" or "turning into" the direction, $\hat{c}$, and taking the "shortest possible path to do so" [10]. This is exactly one way of interpreting a bivector (ref Section B.1).

For a half turn, the two specified direction vectors, $\hat{b}$ and $\hat{c}$, remain nominally pointed in similar directions, or are exactly perpendicular to each other. I.e. for a half-turn, the (classic) angle between the two vectors is never obtuse (it is always either acute or right).

The fact that half-rotations are restricted to non-obtuse angle sizes, eliminates potential ambiguities in rotation expressions (such as the 2-1 mapping ambiguities that occur when using

---

[8]The analogy with "squaring" is actually entirely appropriate. It turns out that the mathematical expression for the rotation operation is indeed a quadratic expression.

[9]This is an operator generalization of the classic "squaring" of a scalar number. E.g. the squared scalar, α^2, can be interpreted as "start with identity (scalar), 1, multiply that first result by α, then multiply the obtained result again by another α". For the rotation case, the description is "start with identity attitude, rotate that first result by a half-rotation, then rotate the obtained result by another half rotation".

[10]E.g. $\hat{b}$ sweeps out a geodesic path on the rotation sphere in the direction of shortest geodesic distance.

unit quaternions or unitary spinors alone to represent rotations). Eliminating this potential for ambiguity provides a solid mathematical foundation that enables powerful mathematical manipulations including the ability to invert many parametric relationships analytically(!) and use familiar algebraic procedures to solve various equations.

It's worth noting another benefit of the half-rotation approach instead of representing the entire rotation with the bivector parameter. The half rotation has a corresponding smaller range, $-\frac{\pi}{2} \leq \frac{\vartheta}{2} \leq \frac{\pi}{2}$. As noted already, the half-rotation is sufficient to represent any possible static attitude uniquely[11]. This particular range of angle size values also provides a powerful practical capability. Specifically, this range of half-angle size values is associated with being able to invert the exponential function without ambiguity[12]. Something that is very useful in theoretical analysis as well as practical computations.

---

[11]For dynamic cases, aliasing or "phase wrapping" considerations need to be addressed. Often that is accomplished with a combination of tracking and counting full cycles and working with incremental rotations each of which is less than a full turn or half turn - i.e. addressed with "book keeping". However, by using bivector angle parameters, the component values can simply be allowed to grow as necessary without concern about restricting their magnitudes (e.g components can have magnitudes that are arbitrarily larger than π.

[12]I.e. is associated with staying on the "principle branch" of the logarithm function.

C. Transformation Concepts

C.1. Convention - Passive vs Active

The transformation convention utilized herein is often called the "passive" convention in which the *coordinate frames* are considered to undergo displacement and turning[1].

The rational for this, is that practical mathematics represent and describe *expressions* of entities *relative to coordinate frames*. By comparison, the physical environment operates directly on the physical object. An easy way to remember this is that: the environment has an active influence on the body, while the mathematics are a passive description of the result.

C.1.1. Illustration Example

As a concrete illustration assume a point of interest, p, has expression in the reference frame, p_R, and $p_R = 2e_1$. If the body is aligned exactly with the reference frame, then the vector expression for this point in the body frame, p_B, will be the same, i.e. $p_B = 2e_1$. Adding a "0" subscript to denote this initial state, the point of interest expressions may be summarized as

$$p_{R,0} = 2e_1$$
$$p_{B,0} = 2e_1$$

---

[1]In contrast to the active convention that is interpreted as the entities (vectors, bivectors) that change state.

If *the body is moved* one unit in the e_1 direction, then the point of interest is expressed after this first change as,

$$p_{R,1} = 2e_1$$
$$p_{B,1} = 1e_1$$

If *the body is then rotated* an eighth of a turn counter clockwise (looking "down" on the e_1e_2 plane), the point of interest expressions after this second change are,

$$p_{R,2} = 2e_1$$
$$p_{B,2} = \frac{1}{\sqrt{2}}e_1 - \frac{1}{\sqrt{2}}e_2$$

C.1.2. General Interpretation

In the foregoing example, the first operation is a displacement *of the body coordinate frame* and the displacement is *followed* by a *rotation of the body coordinate frame*. This is the convention utilized herein. The displacement of the body frame occurs first, and then rotation of the body frame occurs afterward.

The importance of this convention (translate then rotate) is that it allows all parameters to be expressed most simply in the domain of the transformation. The transformation is defined over the domain, the translation (vector) parameters are expressed directly in the domain, and the rotation (bivector) parameters are expressed directly in the domain. I.e. "everything" is expressed in the same domain!!

C.2. Relationships - With Respect To (wrt)

Both the displacement and attitude change can be represented as mathematical transformations that operate on the point of interest vector.

Let $\mathcal{D}_{BwR}$ represent "displacement of the body wrt the reference frame" that is described in first example step above.

$$p_{B,1} = \mathcal{D}_{BwR}\left(p_{R,0}\right)$$

For the second step, let $\mathcal{R}_{BwR}$ represent "rotation of the body wrt the reference frame" such that

$$p_{B,2} - \mathcal{R}_{BwR}\left(p_{B,1}\right)$$

At this point, it's worth a quick interruption to address an important subtlety. The rotation transform has subscripts "BwR", instead of "Bw(Body-after-displacement)". This "subscript hackery" holds up because rotation is independent of translation. Put another way, "rotation does not have an origin", but is a phenomenon that applies throughout the entire body equally. Another way to interpret this is that the "Body-after-displacement" coordinate frame remains exactly and identically aligned with the reference frame - and the subscript hack "glosses over" an implicit intermediate identity transformation.

Since the rotation occurs after the displacement, the rotation operation is applied to the results after having applied the displacement operator, so that

$$p_{B,2} = \mathcal{R}_{BwR}\left(\mathcal{D}_{BwR}\left(p_{R,0}\right)\right)$$

C.3. A Note on Subscript Notations

The description in Section C.2 involves a number of subscripts to help explain the precise mathematical interpretation of the transformation methodology. However, in practice, since all parameters involved on the right side of the transformation relationship are expressed in the same domain, it's common abbreviate considerably.

Since all parameters are expressed in the same domain, it's often practical to use single subscripts in many places and/or change letters to denote different domains (such as the use of "x" and "y" introduced in Chapter 5). However, when extreme precision of expression is required, the use of the "BwA" notation provides a simple notation that is fairly unambiguous.

Compared with some other conventions, the "BwA" notation is less work for typesetting actions since all text goes in one place (e.g. within a single subscript, superscript, under-set, over-set, etc.). The explicit inclusion of the "w" character makes the order of interpretation unambiguous (e.g. contrast with e.g. $*_{CD}$ or $*_{C}^{D}$, which might suggest either one of the opposing interpretations, "from C into D" or "C relative to D").

The "BwA" notation also provides an excellent notation for tracking and interpreting the composition of transformations. E.g. "DwA" can be expressed by the chain of transformations with notation similar to

$$DwA=DwC*CwB*BwA$$

C.4. Canonical Rotation Formula

A completely general rotation operation may be expressed in geometric algebra using only a bivector.

For example, expressing a pure rotation (no translation) that converts a vector's expression, x, into a rotated vector expression, y, may be specified only in terms of the bivector, B, and the exponential function (defined algebraically in Section A.4) as

$$y = e^{\frac{1}{2}B}\left(x\right)e^{-\frac{1}{2}B} \tag{C.4.1}$$

Although not significant here, it is worth noting that this exact same formula applies for rotation in any space in which vectors are defined (e.g. any geometric or vector space with dimension larger than 1). This includes the common practical cases of rotation in a 2D space (e.g. as in image processing), and rotation in 3D space (e.g. situated and moving rigid bodies as being addressed here.

The factor of $\frac{1}{2}$ in the exponents, corresponds with considering the overall rotation as "two half-rotations". The occurrence of two individual exponential factors on the right corresponds with the action of "applying it twice". The sign change in the exponent on the right is associated with a non-commutative difference in behavior between "multiplication from the right" vs "multiplication from the left".

A key point to note here, is that this is a compact rotation formula that is expressed in terms of the minimum required parameters. I.e. for 3D space, the three β_k components of B completely specify rotation (or body attitude) such as in equation A.1.1.

C.5. Spinor (aka Rotor)

For simplicity of notation, it's often convenient to denote and consider the exponential factor, e^H, as an entity in its own right. Let R, be defined as

$$R \equiv e^H$$

The entity, R, is known as a "spinor" or a "rotor". In this context, spinors are isomorphic with the classic quaternions[2].

In GA terms, the spinor, R, is the sum of a scalar grade constituent and a bivector grade constituent. It therefore ap-

---

[2] Although, by comparison, spinors are fundamentally dextral and interact consistently with the rest of the algebra, where quaternions have a slight left-handed "hiccup" in their definition. As an aside, it is this same "hiccup" that historically led to the entirely unnecessary and superfluous distinction between "polar" and "axial" vectors in the classical "Gibbs" style vector algebra formulations.

pears that, R, represents 4 dof (1 in the scalar grade, and 3 in the bivector grade). This apparent paradox is resolved (ref Appendix A.1.6) by noting that the exponential function introduces a constraint, such that

$$|R| = 1$$

Since an arbitrary spinor, R, has four parametric degrees of freedom, it is not suitable to use for rigid body attitude representation unless suitable constraints are introduced. Taking this approach (same with using unit quaternions) injects unnecessary complication into applications - complications such as needing to introduce a separate constraint relationship or needing to "re-normalize" data between iteration steps of numeric computations.

The "reverse" of a spinor[3], is denoted as, $R^\dagger$, and can be defined in terms of the bivector (half) angle, H as

$$R^\dagger = e^{-H}$$

The full rigid body transformation (e.g. relationship 6.0.1) can be expressed in terms of the spinor representation as

$$y = R(x - t)R^\dagger \tag{C.5.1}$$

---

[3] Analogous to the quaternion conjugate in this setting.

D. Exponential Function Derivatives

The derivative of the exponential function may be computed relatively easily.

D.1. Computation for General Case

The derivative of the exponential function may be expressed in terms of the relationship,

$$\left(2H^2\right)\overline{e^H} = \left(H\dot{H} + \dot{H}H\right)He^H - \left(H\dot{H} - \dot{H}H\right)\frac{1}{2}\left(e^H - e^{-H}\right) \tag{D.1.1}$$

For the case where the half-rotation angle is not near zero, its magnitude, $|H|$, is numerically significant, and the leading factor on the left is substantially non-zero (has computational significance). In that case, the derivative can be isolated simply by multiplying both sides with the inverse scalar value, $\left(2H^2\right)^{-1}$.

Values for the exponential function on the right hand side can be evaluated with the formula and techniques described in Section A.4. Note that the exponential of the negative angle is just the reverse of the exponential of the positive angle, i.e.

$$e^{-H} = \left\langle e^H \right\rangle_0 - \left\langle e^H \right\rangle_2$$

For the case of numerically small angles, for which $|H| \leq \varepsilon$ (at some level of machine or application significance, ε), a different computation technique is useful as described in the following.

97

D.2. Computation for Small Values

This section provides a numerically stable computation for the commonly encountered case of evaluating the exponential derivative for small angle values (such cases often occur in numeric difference type computation work).

For the case $|H| \leq \varepsilon$, (with ε determined as appropriate to computational word sizes and/or application precision considerations), the derivative of the exponential function can be evaluated using a series expansion of the exponential function which is appropriately truncated[1].

The general approach is to first specify the desired machine precision, ε_M, then express the exponential function as a series expansion truncated to order M. Next express the $\overline{e^H}$ in terms of the truncated series, then take the limit as $|H| \rightarrow \varepsilon_M$, so that $|H|^M < 0_\varepsilon$, where 0_ε means "numerically insignificant".

For example, to first order, $e^{\pm H} \simeq 1 \pm H$, so that

$$
\left(2H^2\right) \overline{e^H} \simeq \left(H\dot{H} + \dot{H}H\right) H \left(1 + H\right)
$$

$$
- \left(H\dot{H} - \dot{H}H\right) \frac{1}{2} \left((1 + H) - (1 - H)\right)
$$

In neglecting terms of second order and higher, leads to

$$
\left(2H^2\right) \overline{e^H} \simeq 2\left(\dot{H}H\right) H + H\dot{H}H^2 + \dot{H}H^3
$$

$$
\simeq 2\dot{H}H^2 + H\dot{H}H^2 + \dot{H}H^3
$$

$$
\simeq \left(2H^2\right) \left[\dot{H} + \frac{1}{2}H\dot{H} + \frac{1}{2}\dot{H}H\right]
$$

Since both sides scale by the same factor, $2H^2$, the desired

---

[1]The truncation is justified since the exponential series is an alternating series, and therefore the truncation error magnitude is less than the magnitude of the first neglected term.

derivative satisfies

$$\overline{e^H} \simeq \dot{H} + \frac{1}{2}\left(H\dot{H} + \dot{H}H\right) \tag{D.2.1}$$

The second term in the expansion above is the geometric product definition of the (scalar valued) bivector-bivector "dot product" (ref Section A.3), such that the exponential derivative can be expressed, to first order, in terms of the half-angle, H, and its rate of change - the half-angle velocity, $\dot{H}$, as,

$$\overline{e^H} = \dot{H} + \left(H \cdot \dot{H}\right) + \mathcal{O}\left(H^2\right) \tag{D.2.2}$$

This expression provides an efficient means for computation when the $|H|^2 \lesssim \varepsilon_M$ and ε_M is the required precision of the result.

The domain of validity of the series expansion approach can be extended by following the same process, but only neglecting terms after third and higher order.

$$\overline{e^H} = \dot{H} + \left(H \cdot \dot{H}\right)\left(1 + \frac{1}{2}H\right) + \mathcal{O}\left(H^3\right) \tag{D.2.3}$$

This is useful for computation with angle values up to size $|H|^3 \lesssim \varepsilon_M$.

For example, when working with software and a classic "double" value (IEEE 754 64-bit float), the numeric machine precision is on the order of $\varepsilon_M \sim 10^{-15}$. In this case, the second order expansion equation D.2.2 is valid for angle magnitude, $|H|$, up to about 10 nano-radians, while the third order expansion is useful for angle magnitudes up to about 10 micro-radians.

For larger angles, the general expression in equation D.1.1 is reasonably efficient (especially when those exponential functions are evaluated using fast native library and/or firmware support).

D.3. Intuitive Explanation using Small Angles

Motion associated with small angles are a very special case, but also a frequently useful one. The small angle case is also useful for developing some degree of intuition and interpretation of the derivative of the exponential function.

Consider, the fully general rigid body velocity relationship of expression 10.0.1, i.e.,

$$\dot{y} = 2\left\langle \overline{e^H} \left(x - t\right) e^{-H} \right\rangle_1 - e^H \dot{t} e^{-H} + e^H \dot{x} e^{-H}$$

Into this, substitute the derivative relationship D.2.1 for small angle case, $|H| < \varepsilon$, to express the (small angle) velocity relationship as

$$\dot{y} \simeq 2\left\langle \left(\dot{H} + H \cdot \dot{H}\right) \left(x - t\right) e^{-H} \right\rangle_1 - e^H \dot{t} e^{-H} + e^H \dot{x} e^{-H}$$

Introduce the identity relationship

$$e^{-H} e^H = e^H e^{-H} = 1$$

and insert between factors to express the body frame velocity as

$$\dot{y} \simeq 2\left\langle \left(\dot{H} + H \cdot \dot{H}\right) e^{-H} e^H \left(x - t\right) e^{-H} \right\rangle_1 - e^H \dot{t} e^{-H} + e^H \dot{x} e^{-H}$$

In the first term, the last three factors are the same as on the right hand side of the rigid body transform expression 6.0.1. Substitute this later expression to obtain,

$$\dot{y} \simeq 2\left\langle \left(\dot{H} + H \cdot \dot{H}\right) e^{-H} y \right\rangle_1 - e^H \dot{t} e^{-H} + e^H \dot{x} e^{-H} \qquad \text{(D.3.1)}$$

The magnitude of the first term is overall proportional to the size of vector, y. This describes the "lever arm" effect associated with the point of interest, y, being some distance from the turning body's coordinate frame origin[2]. E.g. a point on the rim of a wheel has a faster instantaneous linear speed than a point half way toward the rim.

The other two terms are fairly simple. The second term represents velocity observed in the body frame due to motion of the body itself. E.g. like scenery moving past the window while driving. The third term, captures proper motion of the point of interest. E.g. like observing a change in position of a flying baseball while running around the field.

As the body moves, its translation and attitude change the position of the point of interest in the body frame as represented in the vector, y. As this position changes, the contribution of the first term changes in a complex way that depends on how the attitude is changing, both on the half-angle value and on the half-angle velocity.

This interpretation of the effect of terms is completely general. For the useful general expression refer to relationship 8.2.2. Here, the "small angle" assumption allows interpreting the relationship between rotational and (half-) angular velocities. E.g.

$$\lim_{H \to 0} \Omega \to \left(\dot{H} + H \cdot \dot{H} \right) e^{-H}$$

In practice the fully rigorous expression D.1.1 addresses the entirely general case and should be used for most computations and evaluations with finite (and numerically substantial) attitudes.

D.3.1. Limiting Case $H \to 0$

A further intuition can be gained by considering the case of very small angles for which the limiting case, $H \to 0$, in the series

---

[2]This "lever arm" concept often occurs with various inertial measurement systems in models of the instrumentation equipment payloads.

D. Exponential Function Derivatives

expansions above (e.g. expression D.2.2) yields

$$\overline{e^H} \to \dot{H}$$

Insert this into equation D.3.1 to express the mixed-domain body frame point velocity as,

$$\dot{y} \simeq 2 \left\langle \dot{H} y \right\rangle_1 - e^H \dot{t} e^{-H} + e^H \dot{x} e^{-H}$$

The full angle velocity is twice that of the half-angle velocity, i.e. $\dot{B} = 2\dot{H}$, so that for this special case of small angle attitudes,

$$\dot{y} \simeq \left\langle \dot{B} y \right\rangle_1 - e^H \dot{t} e^{-H} + e^H \dot{x} e^{-H}$$

Comparing this with the corresponding mixed-domain rotational velocity expression 8.2.2 indicates that, for $H \to 0$ (and also $B \to 0$), that the rotational velocity, Ω, and the angular velocity, $\dot{B}$, converge to the same thing. However, this is only true infinitesimally and instantaneously. This *relationship does NOT apply to finite rotations* - not even small finite difference relationships!

Also, note that this limit case is based on the *magnitude* of the half angle, H, being small. The magnitude of the angular velocity, $\left| \dot{H} \right|$, is completely independent of this. Thus, this relationship can be useful in numeric methods. For example, if H is the angular deviation from some prior solution or intermediate computation value, it may have very small magnitude, such that the limit expressions above can be applied, in order to estimate changes in the parameter values associated with solving for, $\dot{H}$, which itself may have a very large magnitude.

This limit case relationship is also exploited in calculus and classical mechanics for making various arguments associated with infinitesimal rotations. However, these idealized limit cases generally do NOT apply to most practical situations such as when working with finite difference methods and discretely sampled kinematic systems! For most all serious practical work, the fully rigorous relationships should be used.

D.4. Exponential Derivatives Recap

The results in first Section D.1 of this appendix are completely general. The second Section D.2 provides fast and numerically stable evaluations for use when the involved angle magnitude is small (an "implementation nuance"). The content of section D.3 is provided primarily to offer and example for developing intuition in understanding the general case expressions.

For practical implementations, use one of the two equation cases to compute the derivative of the exponential function. For larger values of $|H|$, use the general equation D.1.1. For smaller values, use one of the series expansion representations such as equation D.2.2 or D.2.3 which avoid limitations in the precision of numeric computations.

Bibliography

[1] Kent L. Beck. *Extreme programming explained - embrace change*. Addison-Wesley, 2000.

[2] John E. Bortz. A new mathematical formulation for strapdown inertial navigation. *IEEE Trans. Aerospace and Electronic Systems*, Jan 1971. Vol. 1.

[3] Michael Boyle. The integration of angular velocity. *ArXive:1604.08139*, May 2017.

[4] David Hestenes. New foundations for classical mechanics. Kluwer Academic Publishers, 1999.

[5] David Hestenes. The genesis of geometric algebra: A personal retrospective. *Advances in Applied Clifford Algebras*, 27, 03 2017.

[6] J. Lasenby, W.J. Fitzgerald, A.N. Lasenby, and C.J.L. Doran. New geometric methods for computer vision: An application to structure and motion estimation. *International Journal of Computer Vision*, 1998.

[7] Wikipedia contributors. Tait-bryan angles — Wikipedia, the free encyclopedia. https://en.wikipedia.org/w/index.php?title=Tait%E2%80%93Bryan_angles&oldid=393759388, 2010. [Online; accessed 13-November-2021].

[8] Wikipedia contributors. Euler angles — Wikipedia, the free encyclopedia. https://en.wikipedia.org/w/index.

`php?title=Euler_angles&oldid=1054152729`, 2021. [Online; accessed 13-November-2021].

[9] Wikipedia contributors. Euler's rotation theorem — Wikipedia, the free encyclopedia. `https://en.wikipedia.org/w/index.php?title=Euler%27s_rotation_theorem&oldid=1050406121`, 2021. [Online; accessed 12-November-2021].

[10] Wikipedia contributors. Rigid body dynamics — Wikipedia, the free encyclopedia. `https://en.wikipedia.org/w/index.php?title=Rigid_body_dynamics&oldid=1061290978`, 2021. [Online; accessed 24-May-2022].

[11] Wikipedia contributors. Euler's formula — Wikipedia, the free encyclopedia. `https://en.wikipedia.org/w/index.php?title=Euler%27s_formula&oldid=1088782685`, 2022. [Online; accessed 29-May-2022].

[12] Wikipedia contributors. Gimbal lock — Wikipedia, the free encyclopedia. `https://en.wikipedia.org/w/index.php?title=Gimbal_lock&oldid=1082887600`, 2022. [Online; accessed 26-May-2022].

[13] Wikipedia contributors. Slerp — Wikipedia, the free encyclopedia. `https://en.wikipedia.org/w/index.php?title=Slerp&oldid=1082816849`, 2022. [Online; accessed 26-May-2022].

Index

Author Biography

Dave Knopp provides algorithm, math model, and software development services for Stellacore Corporation. He has first hand experience implementing the approaches and techniques described herein and has realized great success in doing so across many disciplines and applications.

Book Jacket

This book offers a practical and concise formulation of the geometry and mathematics associated with position and attitude of a rigid body in 3D space. The material is presented as a practical technical synopsis containing useful formulas and expressions presented in context of general descriptions and underlying physical interpretations. This material provides a pragmatic "bootstrap" style resource for those working with the concepts of rigid body position, attitude and motion. The mathematical model offers an extremely efficient, and exceptionally effective framework for practical implementation. The content is structured as a useful working-reference intended for those who are developing algorithms and applications involved with modeling, measuring and/or controlling rigid objects in our physical world. Readers are assumed to have general math skills that cover classic algebras and basic calculus.

www.ingramcontent.com/pod-product-compliance
Lightning Source LLC
Chambersburg PA
CBHW071228170526
45165CB00003B/1039